SCIENCE IN ACTION

SCIENCE IN ACTION

How to follow
scientists and engineers
through society

Bruno Latour

Harvard University Press
Cambridge, Massachusetts

Library of Congress Cataloging in Publication Data
Main entry under title:

Latour, Bruno.
  Science in action.

  Bibliography: p.
  Includes index.
  1. Science—Social aspects.   2. Technology—Social aspects.
  3. Science—History.   4. Science—Philosophy.
I. Title.
Q175.5.L38 1987 306'.45   86-16326

ISBN 0-674-79291-2 (paper)

To Michel Callon,
this outcome of a seven-year discussion

Contents

Acknowledgements

Not being a native English speaker I had to rely heavily on my friends to revise successive drafts of this manuscript. John Law and Penelope Dulling have been most patient in revising the earlier drafts. Steven Shapin, Harry Collins, Don MacKenzie, Ron Westrum and Leigh Star suffered each on one different chapter. I have been most fortunate in having Geoffrey Bowker edit the whole book, 'debug' it and suggest many useful changes.

Part of the work for this book has been supported by a grant from CNRS-*Programme STS*. Not a line of it could have been written without the stimulation, environment, friendship and material conditions of the Centre de Sociologie de l'Innovation at Ecole Nationale Supérieure des Mines de Paris, my new 'alma mater'.

INTRODUCTION

Opening Pandora's Black Box

Scene 1: On a cold and sunny morning in October 1985, John Whittaker entered his office in the molecular biology building of the Institut Pasteur in Paris and switched on his *Eclipse MV/8000* computer. A few seconds after loading the special programs he had written, a three-dimensional picture of the DNA double helix flashed onto the screen. John, a visiting computer scientist, had been invited by the Institute to write programs that could produce three-dimensional images of the coils of DNA and relate them to the thousands of new nucleic acid sequences pouring out every year into the journals and data banks. 'Nice picture, eh?' said his boss, Pierre, who was just entering the office. 'Yes, good machine too,' answered John.

Scene 2: In 1951 in the Cavendish laboratory at Cambridge, England, the X-ray pictures of crystallised deoxyribonucleic acid were not 'nice pictures' on a computer screen. The two young researchers, Jim Watson and Francis Crick[1], had a hard time obtaining them from Maurice Wilkins and Rosalind Franklin in London. It was impossible yet to decide if the form of the acid was a triple or a double helix, if the phosphate bonds were at the inside or at the outside of the molecule, or indeed if it was an helix at all. It did not matter much to their boss, Sir Lawrence Bragg, since the two were not supposed to be working on DNA anyway, but it mattered a lot to them, especially since Linus Pauling, the famous chemist, was said to be about to uncover the structure of DNA in a few months.

Scene 3: In 1980 in a Data General building on Route 495 in Westborough, Massachusetts, Tom West[2] and his team were still trying to debug a makeshift prototype of a new machine nicknamed *Eagle* that the company had not planned to build at first, but that was beginning to rouse the marketing department's interest. However, the debugging program was a year behind schedule. Besides, the choice West had made of using the new PAL chips kept delaying the machine – renamed *Eclipse MV/8000*, since no one was sure at the time if the company manufacturing the chips could deliver them on demand. In the meantime, their main competitor, DEC, was selling many copies of its *VAX 11/780*, increasing the gap between the two companies.

(1) *Looking for a way in*

Where can we start a study of science and technology? The choice of a way in crucially depends on good timing. In 1985, in Paris, John Whittaker obtains 'nice pictures' of DNA on a 'good machine'. In 1951 in Cambridge Watson and Crick are struggling to define a shape for DNA that is compatible with the pictures they glimpsed in Wilkins's office. In 1980, in the basement of a building, another team of researchers is fighting to make a new computer work and to catch up with DEC. What is the meaning of these 'flashbacks', to use the cinema term? They carry us back through space and time.

When we use this travel machine, DNA ceases to have a shape so well established that computer programs can be written to display it on a screen. As to the computers, they don't exist at all. Hundreds of nucleic acid sequences are not pouring in every year. Not a single one is known and even the notion of a sequence is doubtful since it is still unsure, for many people at the time, whether DNA plays any significant role in passing genetic material from one generation to the next. Twice already, Watson and Crick had proudly announced that they had solved the riddle and both times their model had been reduced to ashes. As to the 'good machine' *Eagle*, the flashback takes us back to a moment when it cannot run any program at all. Instead of a routine piece of equipment John Whittaker can switch on, it is a disorderly array of cables and chips surveyed by two other computers and surrounded by dozens of engineers trying to make it work reliably for more than a few seconds. No one in the team knows yet if this project is not going to turn out to be another complete failure like the *EGO* computer on which they worked for years and which was killed, they say, by the management.

In Whittaker's research project many things are unsettled. He does not know how long he is going to stay, if his fellowship will be renewed, if any program of his own can handle millions of base pairs and compare them in a way that is biologically significant. But there are at least two elements that raise no problems for him: the double helix shape of DNA and his Data General computer. What was for Watson and Crick the problematic focus of a fierce challenge, what won them a Nobel Prize, is now the basic dogma of his program, embedded in thousand of lines of his listing. As for the machine that made West's team work day and night for years, it is now no more problematic than a piece of furniture as it hums quietly away in his office. To be sure, the maintenance man of Data General stops by every week to fix up some minor problems; but neither the man nor John have to overhaul the computer all over again and force the company to develop a new line of products. Whittaker is equally well aware of the many problems plaguing the Basic Dogma of biology – Crick, now an old gentleman, gave a lecture at the Institute on this a few weeks ago – but neither John nor his boss have to rethink entirely the shape of the double helix or to establish a new dogma.

The word **black box** is used by cyberneticians whenever a piece of machinery or a set of commands is too complex. In its place they draw a little box about which they need to know nothing but its input and output. As far as John Whittaker is concerned the double helix and the machine are two black boxes. That is, no matter how controversial their history, how complex their inner workings, how large the commercial or academic networks that hold them in place, only their input and output count. When you switch on the *Eclipse* it runs the programs you load; when you compare nucleic acid sequences you start from the double helix shape.

The flashback from October 1985 in Paris to Autumn 1951 in Cambridge or December 1980 in Westborough, Massachusetts, presents two completely different pictures of each of these two objects, a scientific fact – the double-helix – and a technical artefact – the *Eagle* minicomputer. In the first picture John Whittaker uses two black boxes because they are unproblematic and certain; during the flashback the boxes get reopened and a bright coloured light illuminates them. In the first picture, there is no longer any need to decide where to put the phosphate backbone of the double helix, it is just there at the outside; there is no longer any squabble to decide if the *Eclipse* should be a 32-bit fully compatible machine, as you just hook it up to the other NOVA computers. During the flashbacks, a lot of people are introduced back into the picture, many of them staking their career on the *decisions* they take: Rosalind Franklin decides to reject the model-building approach Jim and Francis have chosen and to concentrate instead on basic X-ray crystallography in order to obtain better photographs; West decides to make a 32-bit compatible machine even though this means building a tinkered 'kludge', as they contemptuously say, and losing some of his best engineers, who want to design a neat new one.

In the Pasteur Institute John Whittaker is taking no big risk in believing the three-dimensional shape of the double helix or in running his program on the *Eclipse*. These are now routine choices. The risks he and his boss take lie elsewhere, in this gigantic program of comparing all the base pairs generated by molecular biologists all over the world. But if we go back to Cambridge, thirty years ago, who should we believe? Rosalind Franklin who says it might be a three-strand helix? Bragg who orders Watson and Crick to give up this hopeless work entirely and get back to serious business? Pauling, the best chemist in the world, who unveils a structure that breaks all the known laws of chemistry? The same uncertainty arises in the Westborough of a few years ago. Should West obey his boss, de Castro, when he is explicitly asked *not* to do a new research project there, since all the company research has now moved to North Carolina? How long should West pretend he is not working on a new computer? Should he believe the marketing experts when they say that all their customers want a fully compatible machine (on which they can reuse their old software) instead of doing as his competitor DEC does a 'culturally compatible' one (on which they cannot reuse their software but only the most basic commands)? What confidence should he have in his old team burned out by the failure of the *EGO* project? Should he risk using the new PAL chips instead of the older but safer ones?

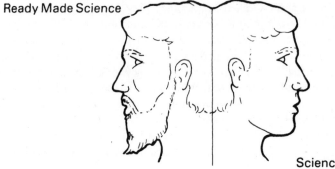

Ready Made Science

Science in the Making

Figure I.1

Uncertainty, people at work, decisions, competition, controversies are what one gets when making a flashback from certain, cold, unproblematic black boxes to their recent past. If you take two pictures, one of the black boxes and the other of the open controversies, they are utterly different. They are as different as the two sides, one lively, the other severe, of a two-faced Janus. 'Science in the making' on the right side, 'all made science' or 'ready made science' on the other; such is Janus *bifrons*, the first character that greets us at the beginning of our journey.

In John's office, the two black boxes cannot and should not be reopened. As to the two controverial pieces of work going on in the Cavendish and in Westborough, they are laid open for us by the scientists at work. The impossible task of opening the black box is made feasible (if not easy) by moving in time and space until one finds the controversial topic on which scientists and engineers are busy at work. This is the first decision we have to make: our entry into science and technology will be through the back door of science in the making, not through the more grandiose entrance of ready made science.

Now that the way in has been decided upon, with what sort of prior knowledge should one be equipped before entering science and technology? In John Whittaker's office the double helix model and the computer are clearly distinct from the rest of his worries. They do not interfere with his psychological mood, the financial problems of the Institute, the big grants for which his boss has applied, or with the political struggle they are all engaged in to create in France a big data bank for molecular biologists. They are just sitting there in the background, their scientific or technical contents neatly distinct from the mess that John is immersed in. If he wishes to know something about the DNA structure or about the *Eclipse*, John opens *Molecular Biology of the Gene* or the *User's Manual*, books that he can take off the shelf. However, if we go back to Westborough or to Cambridge this clean distinction between a context and a content disappears.

Scene 4: Tom West sneaks into the basement of a building where a friend lets him in at night to look at a *VAX* computer. West starts pulling out the printed circuits boards and analyses his competitor. Even his first analysis merges technical and quick economic calculations with the strategic decisions already taken. After a few hours, he is reassured.

> 'I'd been living in fear of VAX for a year,' West said afterward. (. . .) 'I think I got a high when I looked at it and saw how complex and expensive it was. It made me feel good about some of the decisions we've made'.

Then his evaluation becomes still more complex, including social, stylistic and organisational features:

> Looking into the VAX, West had imagined he saw a diagram of DEC's corporate organization. He felt that VAX was too complicated. He did not like, for instance, the system by which various parts of the machine communicated with each other, for his taste, there was too much protocol involved. He decided that VAX embodied flaws in DEC's corporate organization. The machine expressed that phenomenally successful company's cautious, bureaucratic style. Was this true? West said it did not matter, it was a useful theory. Then he rephrased his opinions. 'With VAX, DEC was trying to minimize the risk', he said, as he swerved around another car. Grinning, he went on: 'We're trying to maximize the win, and make Eagle go as fast as a raped ape.'
>
> (Kidder: 1981, p. 36)

This heterogeneous evaluation of his competitor is not a marginal moment in the story; it is the crucial episode when West decides that in spite of a two-year delay, the opposition of the North Carolina group, the failure of the *EGO* project, they can still make the *Eagle* work. 'Organisation', 'taste', 'protocol', 'bureaucracy', 'minimisation of risks', are not common technical words to describe a chip. This is true, however, only once the chip is a black box sold to consumers. When it is submitted to a competitor's trial, like the one West does, all these bizarre words become part and parcel of the technical evaluation. Context and contents merge.

Scene 5: Jim Watson and Francis Crick get a copy of the paper unveiling the structure of DNA written by Linus Pauling and brought to them by his son:

> Peter's face betrayed something important as he entered the door, and my stomach sank in apprehension at learning that all was lost. Seeing that neither Francis nor I could bear any further suspense, he quickly told us that the model was a three-chain helix with the sugar phosphate backbone in the center. This sounded so suspiciously like our aborted effort of last year that immediately I wondered whether we might already have had the credit and glory of a great discovery if Bragg had not held us back.
>
> (Watson: 1968, p. 102)

Was it Bragg who made them miss a major discovery, or was it Linus who missed a good opportunity for keeping his mouth shut? Francis and Jim hurriedly try out the paper and look to see if the sugar phosphate backbone is solid enough to hold the structure together. To their amazement, the three chains described by Pauling had

no hydrogen atoms to tie the three strands together. Without them, if they knew
their chemistry, the structure will immediately fly apart.

> Yet somehow Linus, unquestionably the world's most astute chemist, had
> come to the opposite conclusion. When Francis was amazed equally by
> Pauling's unorthodox chemistry, I began to breathe slower. By then I knew
> we were still in the game. Neither of us, however, had the slightest clue to the
> steps that had led Linus to his blunder. If a student had made a similar
> mistake, he would be thought unfit to benefit from Cal Tech's chemistry
> faculty. Thus, we could not but initially worry whether Linus's model
> followed from a revolutionary reevaluation of the acid-based properties of
> very large molecules. The tone of the manuscript, however, argued against
> any such advance in chemical theory.
>
>   (idem: p. 103)

To decide whether they are still in the game Watson and Crick have to
evaluate simultaneously Linus Pauling's reputation, common chemistry, the
tone of the paper, the level of Cal Tech's students; they have to decide if a
revolution is under way, in which case they have been beaten off, or if an
enormous blunder has been committed, in which case they have to rush still faster
because Pauling will not be long in picking it up:

> When his mistake became known, Linus would not stop until he had captured the
> right structure. Now our immediate hope was that his chemical colleagues would be
> more than ever awed by his intellect and not probe the details of his model. But since
> the manuscript had already been dispatched to the *Proceedings of the National
> Academy*, by mid-March at the latest Linus's paper would be spread around the
> world. Then it would be only a matter of days before the error would be discovered.
> We had anywhere up to six weeks before Linus again was in full-time pursuit of
> DNA.
>
>   (idem: p. 104)

'Suspense', 'game', 'tone', 'delay of publication', 'awe', 'six weeks delay' are
not common words for describing a molecule structure. This is the case at least
once the structure is known and learned by every student. However, as long as the
structure is submitted to a competitor's probing, these queer words are part and
parcel of the very chemical structure under investigation. Here again context and
content fuse together.

The equipment necessary to travel through science and technology is at once
light and multiple. Multiple because it means mixing hydrogen bonds with
deadlines, the probing of one another's authority with money, debugging and
bureaucratic style; but the equipment is also light because it means simply leaving
aside all the prejudices about what distinguishes the context in which knowledge
is embedded and this knowledge itself. At the entrance of Dante's Inferno is
written:

ABANDON HOPE ALL YE WHO ENTER HERE.

At the onset of this voyage should be written:

ABANDON KNOWLEDGE ABOUT KNOWLEDGE
ALL YE WHO ENTER HERE.

Learning to use the double helix and *Eagle* in 1985 to write programs reveals none of the bizarre mixture they are composed of; studying these in 1952 or in 1980 reveals it all. On the two black boxes sitting in Whittaker's office it is inscribed, as on Pandora's box: DANGER: DO NOT OPEN. From the two tasks at hand in the Cavendish and in Data General Headquarters, passions, deadlines, decisions escape in all directions from a box that lies open. Pandora, the mythical android sent by Zeus to Prometheus, is the second character after Janus to greet us at the beginning of our trip. (We might need more than one blessing from more than one of the antique gods if we want to reach our destination safely.)

(2) *When enough is never enough*

Science has two faces: one that knows, the other that does not know yet. We will choose the more ignorant. Insiders, and outsiders as well, have lots of ideas about the ingredients necessary for science in the making. We will have as few ideas as possible on what constitutes science. But how are we going to account for the closing of the boxes, because they do, after all, close up? The shape of the double helix is settled in John's office in 1985; so is that of the *Eclipse MV/8000* computer. How did they move from the Cavendish in 1952 or from Westborough, Massachusetts, to Paris 1985? It is all very well to choose controversies as a way in, but we need to follow also the closure of these controversies. Here we have to get used to a strange acoustic phenomenon. The two faces of Janus talk at once and they say entirely different things that we should not confuse.

Janus' first dictum:

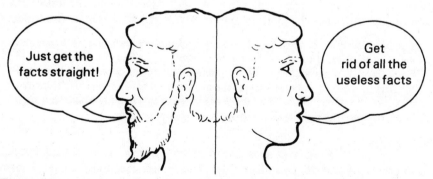

Figure I.2

Scene 6: Jim copies from various textbooks the forms of the base pairs that make up DNA, and plays with them trying to see if a symmetry can be seen when pairing them. To his amazement adenine coupled with adenine, cytosine with cytosine, guanine with guanine and thymine with thymine make very nice superimposable forms. To be sure this symmetry renders the sugar phosphate backbone strangely misshapen but this is not enough to stop Jim's pulse racing or to stop him writing a triumphant letter to his boss.

> I no sooner got to the office and began explaining my scheme than the American crystallographer Jerry Donohue protested that the idea would not work. The tautomeric forms I had copied out of Davidson's book were, in Jerry's opinion, incorrectly assigned. My immediate retort that several other texts also pictured guanine and thymine in the enol form cut no ice with Jerry. Happily he let out that for years organic chemists had been arbitrarily favoring particular tautomeric forms over their alternatives on only the flimsiest of grounds. (. . .) Though my immediate reaction was to hope that Jerry was blowing hot air, I did not dismiss his criticism. Next to Linus himself, Jerry knew more about hydrogen bonds than anyone in the world. Since for many years he had worked at Cal Tech on the crystal structures of small organic molecules, I couldn't kid myself that he did not grasp our problem. During the six months that he occupied a desk in our office, I had never heard him shooting off his mouth on subjects about which he knew nothing. Thoroughly worried, I went back to my desk hoping that some gimmick might emerge to salvage the like-with-like idea.
>
> (Watson: 1968, pp. 121–2)

Jim had got the facts straight out of textbooks which, unanimously, provided him with a nice black box: the enol form. In this case, however, this is the very fact that should be dismissed or put into question. Or at least this is what Donohue says. But whom should Jim believe? The unanimous opinion of organic chemists or *this* chemist's opinion? Jim, who tries to salvage his model, switches from one rule of method, 'get the facts straight', to other more strategic ones, 'look for a weak point', 'choose who to believe'. Donohue studied with Pauling, he worked on small molecules, in six months he never said absurd things. Discipline, affiliation, curriculum vitae, psychological appraisal are mixed together by Jim to reach a decision. Better sacrifice them and the nice like-with-like model, than Donohue's criticism. The fact, no matter how 'straight', has to be dismissed.

> The unforeseen dividend of having Jerry share an office with Francis, Peter, and me, though obvious to all, was not spoken about. If he had not been with us in Cambridge, I might still have been pumping out for a like-with-like structure. Maurice, in a lab devoid of structural chemists, did not have anyone to tell him that all the textbook pictures were wrong. But for Jerry, only Pauling would have been likely to make the right choice and stick by its consequences.
>
> (idem: p. 132)

The advice of Janus' left side is easy to follow when things are settled, but not as long as things remain unsettled. What is on the left side, universal well-known facts of chemistry, becomes, from the right side point of view, scarce pronouncements uttered by two people in the whole world. They have a *quality* that crucially depends on localisation, on chance, on appraising simultaneously the worth of the people and of what they say.

Janus's second dictum:

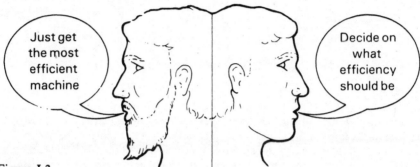

Figure I.3

Scene 7: West and his main collaborator, Alsing, are discussing how to tackle the debugging program:

'I want to build a simulator, Tom.'
'It'll take too long, Alsing. The machine'll be debugged before you get your simulator debugged.'
This time, Alsing insisted. They could not build Eagle in anything like a year if they had to debug all the microcode on prototypes. If they went that way, moreover, they'd need to have at least one and probably two extra prototypes right from the start, and that would mean a doubling of the boring, grueling work of updating boards. Alsing wanted a program that would behave like a perfected Eagle, so that they could debug their microcode separately from the hardware.
West said: 'Go ahead. But I betchya it'll all be over by the time you get it done.'

(Kidder: 1981, p. 146)

The right side's advice is strictly followed by the two men since they want to build the best possible computer. This however does not prevent a new controversy starting between the two men on how to mimic in advance an efficient machine. If Alsing cannot convince one of his team members, Peck, to finish in six weeks the simulator that should have taken a year and a half, then West will be right: the simulator is not an efficient way to proceed because it will come too late. But if Alsing and Peck succeed, then it is West's definition of efficiency which will turn out to be wrong. Efficiency will be the consequence of who succeeds; it does not help deciding, on the spot, who is right and wrong. The right side's advice is all very well once *Eagle* is sent to manufacturing; before that, it is the left side's confusing strategic advice that should be followed.

Janus' third dictum:

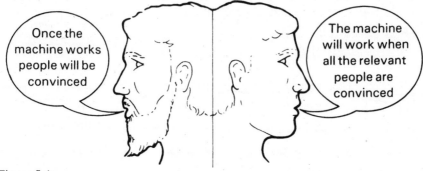

Once the machine works people will be convinced

The machine will work when all the relevant people are convinced

Figure I.4

Scene 8: West has insulated his team for two years from the rest of the company. 'Some of the kids,' he says, 'don't have a notion that there's a company behind all of this. It could be the CIA funding this. It could be a psychological test' (Kidder: 1982, p. 200). During this time, however, West has constantly lobbied the company on behalf of *Eagle*. Acting as a middle-man he has filtered the constraints imposed on the future machine by de Castro (the Big Boss), the marketing department, the other research group in North Carolina, the other machines presented in computer fairs, and so on. He was also the one who kept negotiating the deadlines that were never met. But there comes a point when all the other departments he has lobbied so intensely want to see something, and call his bluff. The situation becomes especially tricky when it is clear at last that the North Carolina group will not deliver a machine, that DEC is selling *VAX* like hot cakes and that all the customers want a supermini 32-bit fully compatible machine from Data General. At this point West has to break the protective shell he has built around his team. To be sure, he designed the machine so as to fit it in with the other departments' interests, but he is still uncertain of their reaction and of that of his team suddenly bereft of the machine.

> As the summer came on, increasing numbers of intruders were being led into the lab – diagnostic programmers and, particularly, those programmers from Software. Some Hardy Boys had grown fond of the prototypes of Eagle, as you might of a pet or a plant you've raised from a seedling. Now Rasala was telling them that they couldn't work on their machines at certain hours, because Software needed to use them. There was an explanation: the project was at a precarious stage; if Software didn't get to know and like the hardware and did not speak enthusiastically about it, the project might be ruined; the Hardy Boys were lucky that Software wanted to use the prototypes – and they had to keep Software happy.
>
> (idem: p. 201)

Not only the Software people have to be kept happy, but also the manufacturing people, those from marketing, those who write the technical documentation, the designers who have to place the whole machine in a nice looking box (not a black one this time!), not mentioning the stockholders and the customers. Although the machine has been conceived by West, through many compromises, to keep all these people happy and busy, he cannot be sure it is going to hold them together. Each of the interest groups has to try their own different sort of tests on the machine and see how it withstands them. The worst, for Tom West, is that the company manufacturing the new PAL chips is going bankrupt, that the team is suffering a *post partum* depression, and that the machine is not yet debugged. 'Our credibility, I think, is running out,' West tells his assistants. *Eagle* still does not run more than a few seconds without flashing error messages on the screen. Every time they painstakingly pinpoint the bug, they fix it and then try a new and more difficult debugging program.

> Eagle was failing its Multiprogramming Reliability Test mysteriously. It was blowing away, crashing, going out to never-never land, and falling off the end of the world after every four hours or so of smooth running.
> 'Machines somewhere in the agony of the last few bugs are very vulnerable,' says Alsing. 'The shouting starts about it. It'll never work, and so on. Managers and support groups start saying this. Hangers-on say, "Gee, I thought you'd get it done a lot sooner." That's when people start talking about redesigning the whole thing.'
> Alsing added, 'Watch out for Tom now.'
> West sat in his office. 'I'm thinking of throwing the kids out of the lab and going in there with Rasala and fix it. It's true. I don't understand all the details of that sucker, but I will, and I'll get it to work.'
> 'Gimme a few more days,' said Rasala.
>
> (idem: p. 231)

A few weeks later, after *Eagle* has successfully run a computer game called Adventure, the whole team felt they had reached one approximate end: 'It's a computer,' Rasala said (idem: p. 233). On Monday 8 October, a maintenance crew comes to wheel down the hall what was quickly becoming a black box. Why has it become such? Because it is a good machine, says the left side of our Janus friend. But it was not a good machine before it worked. Thus while it is being made it cannot convince anyone *because* of its good working order. It is only after endless little bugs have been taken out, each bug being revealed by a new trial imposed by a new interested group, that the machine will *eventually* and *progressively* be made to work. All the reasons for why it will work once it is finished do not help the engineers while they are making it.

Scene 9: How does the double helix story end? In a series of trials imposed on the new model by each of the successive people Jim Watson and Francis Crick have worked with (or against). Jim is playing with cardboard models of the base pairs, now in the keto form suggested by Jerry Donohue. To his amazement he realises that the shape drawn by pairing adenine with thymine and guanine with cytosine are superimposable. The steps of the double helix have the same shape. Contrary to his earlier model, the structure might be complementary instead of being like-with-like. He hesitates a while, because he sees no reason at first for this complementarity. Then he remembers what was called 'Chargaff laws', one of these many empirical facts they had kept in the background. These 'laws' stated that there

Janus's fourth dictum:

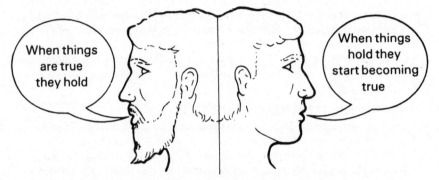

When things are true they hold

When things hold they start becoming true

was always as much adenine as thymine and as much guanine as cytosine, no matter which DNA one chose to analyse. This isolated fact, devoid of any meaning in his earlier like-with-like model, suddenly brings a new strength to his emerging new model. Not only are the pairs superimposable, but Chargaff laws can be made a consequence of his model. Another feature came to strengthen the model: it suggests a way for a gene to split into two parts and then for each strand to create an exact complementary copy of itself. One helix could give birth to two identical helices. Thus biological meaning could support the model.

Still Jim's cardboard model could be destroyed in spite of these three advantages. Maybe Donohue will burn it to ashes as he did the attempt a few days earlier. So Jim called him to check if he had any objection. 'When he said no, my morale skyrocketed' (Watson: 1968, p. 124). Then it is Francis who rushes into the lab and 'pushes the bases together in a number of ways'. The model, this time, *resists* Francis's scepticism. There are now many decisive elements tied together with and by the new structure.

Still, all the convinced people are in the same office and although they think they are right, they could still be deluding themselves. What will Bragg and all the other crystallographers say? What objections will Maurice Wilkins and Rosalind Franklin, the only ones with X-rays pictures of the DNA, have? Will they see the model as *the* only form able to give, by projection, the shape visible on Rosalind's photographs? They'd like to know fast but dread the danger of the final showdown with people who, several times already, have ruined their efforts. Besides, another ally is missing to set up the trial, a humble ally for sure but necessary all the same: 'That night, however, we could not firmly establish the double helix. Until the metal bases were on hand, any model building would be too sloppy to be convincing' (idem: p. 127). Even with Chargaff laws, with biological significance, with Donohue's approval, with their excitement, with the base pairing all on their side, the helix is still sloppy. Metal is necessary to reinforce the structure long enough to withstand the trials that the competitors/colleagues are going to impose on it.

The remainder of the double helix story looks like the final rounds of a presidential nomination. Every one of the other contenders is introduced into the office where the model is now set up, fights with it for a while before being quickly overwhelmed and then pledging complete support to it. Bragg is convinced although still worried that no one more serious than Jim and Francis had checked the helix. Now for the big game, the encounter between the model and those who for years had captured its projected image. 'Maurice needed but a minute's look at the model to like it.' 'He was back in London only two days before he rang up to say that both he and Rosy found that their X-ray data strongly supported the double helix' (p. 131). Soon Pauling rallies himself to the structure, then it is the turn of the referees of *Nature*.

'Of course,' says the left side of Janus, 'everyone is convinced because Jim and Francis stumbled on the right structure. The DNA shape itself is enough to rally everyone.' 'No, says the right side, every time someone else is convinced it progressively becomes a more right structure.' Enough is never enough: years later in India and New Zealand other researchers were working on a so-called 'warped zipper'[3] model that did everything the double helix does – plus a bit more; Pauling strongly supported his own structure that had turned out to be entirely wrong; Jim found biological significance in a like-with-like structure that survived only a few hours; Rosalind Franklin had been stubbornly convinced earlier that it was a three-strand helix; Wilkins ignored the keto forms revealed by Jerry Donohue; Chargaff's laws were an insignificant fact they kept in the background for a long time; as to the metal atom toys, they have lent strong support to countless models that turned out to be wrong. All these allies appear strong once the structure is blackboxed. As long as it is not, Jim and Francis are still struggling to recruit them, modifying the DNA structure until everyone is satisfied. When they are through, they will follow the advice of Janus's right side. As long as they are still searching for the right DNA shape, they would be better off following the right side's confusing advices.

We could review all the opinions offered to explain why an open controversy closes, but we will always stumble on a new controversy dealing with how and why it closed. We will have to learn to live with two contradictory voices talking at once, one about science in the making, the other about ready made science. The latter produces sentences like 'just do this . . . just do that . . .'; the former says 'enough is never enough'. The left side considers that facts and machines are well determined enough. The right side considers that facts and machines in the making are always **under-determined.**[4] Some little thing is always missing to close the black box once and for all. Until the last minute *Eagle* can fail if West is not careful enough to keep the Software people interested, to maintain the pressure on the debugging crew, to advertise the machine to the marketing department.

(3) The first rule of method

We will enter facts and machines while they are in the making; we will carry with us no preconceptions of what constitutes knowledge; we will watch the closure of

Figure I.6

the black boxes and be careful to distinguish between two contradictory explanations of this closure, one uttered when it is finished, the other while it is being attempted. This will constitute our **first rule of method** and will make our voyage possible.

To sketch the general shape of this book, it is best to picture the following comic strip: we start with a textbook sentence which is devoid of any trace of fabrication, construction or ownership; we then put it in quotation marks, surround it with a bubble, place it in the mouth of someone who speaks; then we add to this speaking character another character *to whom* it is speaking; then we place all of them in a specific situation, somewhere in time and space, surrounded by equipment, machines, colleagues; then when the controversy heats up a bit we look at *where* the disputing people go and *what* sort of new elements they fetch, recruit or seduce in order to convince their colleagues; then, we see how the people being convinced stop discussing with one another; situations, localisations, even people start being slowly erased; on the last picture we see a new sentence, without any quotation marks, written in a text book similar to the one we started with in the first picture. This is the general movement of what we will study over and over again in the course of this book, penetrating science from the outside, following controversies and accompanying scientists up to the end, being slowly led out of science in the making.

In spite of the rich, confusing, ambiguous and fascinating picture that is thus revealed, surprisingly few people have penetrated from the outside the inner workings of science and technology, and then got out of it to explain to the outsider how it all works. For sure, many young people have entered science, but they have become scientists and engineers; what they have done is visible in the machines we use, the textbooks we learn, the pills we take, the landscape we look at, the blinking satellites in the night sky above our head. How they did it, we don't know. Some scientists talk about science, its ways and means, but few of them accept the discipline of becoming also an outsider; what they say about their trade is hard to double check in the absence of independent scrutiny. Other people talk about science, its solidity, its foundation, its development or its dangers; unfortunately, almost none of them are interested in science in the making. They shy away from the disorderly mixture revealed by science in action and prefer the orderly pattern of scientific method and rationality. Defending science and reason against pseudo-sciences, against fraud, against irrationality, keeps most of these people too busy to study it. As to the millions, or billions, of outsiders, they know about science and technology through popularisation only. The facts and the artefacts they produce fall on their head like an external fate as foreign, as inhuman, as unpredictable as the olden *Fatum* of the Romans.

Apart from those who make science, who study it, who defend it or who submit to it, there exist, fortunately, a few people, either trained as scientists or not, who open the black boxes so that outsiders may have a glimpse at it. They go by many different names (historians of science and technology, economists, sociologists, science teachers, science policy analysts, journalists, philosophers, concerned scientists and citizens, cognitive anthropologists or cognitive psychologists), and are most often filed under the general label of 'science, technology and society'. It is on their work that this book is built. A summary of their many *results* and achievements would be worth doing, but is beyond the scope of my knowledge. I simply wish to summarise their *method* and to sketch the ground that, sometimes unwittingly, they all have in common. In doing so I wish to help overcome two of the limitations of 'science, technology and society' studies that appear to me to thwart their impact, that is their organisation *by discipline* and *by object*.

Economists of innovation ignore sociologists of technology; cognitive scientists never use social studies of science; ethnoscience is far remote from pedagogy; historians of science pay little attention to literary studies or to rhetoric; sociologists of science often see no relation between their academic work and the *in vivo* experiments performed by concerned scientists or citizens; journalists rarely quote scholarly work on social studies of science; and so on.

This Babel of disciplines would not matter much if it was not worsened by another division made according to the objects each of them study. There exist historians of eighteenth-century chemistry or of German turn-of-the-century physics; even citizens' associations are specialised, some in fighting atomic energy, others in struggling against drug companies, still others against new math teaching; some cognitive scientists study young children in experimental settings while others are interested in adult daily reasoning; even among sociologists of science, some focus on micro-studies of science while others tackle large-scale engineering projects; historians of technology are often aligned along the technical specialities of the engineers, some studying aircraft industries while others prefer telecommunications or the development of steam engines; as to the anthropologists studying 'savage' reasoning, very few get to deal with modern knowledge. This scattering of disciplines and objects would not be a problem if it was the hallmark of a necessary and fecund *specialisation*, growing from a core of common problems and methods. This is however far from the case. The sciences and the technologies to be studied are the main factors in determining this haphazard growth of interests and methods. I have never met two people who could agree on what the domain called 'science, technology and society' meant – in fact, I have rarely seen anyone agree on the name or indeed that the domain exists!

I claim that the domain exists, that there is a core of common problems and methods, that it is important and that all the disciplines and objects of 'science, technology and society' studies can be employed as so much specialised material with which to study it. To define what is at stake in this domain, the only thing we need is a few sets of concepts sturdy enough to stand the trip through all these many disciplines, periods and objects.

I am well aware that there exist many more sophisticated, subtle, fast or powerful notions than the ones I have chosen. Are they not going to break down? Are they going to last the distance? Will they be able to tie together enough empirical facts? Are they handy enough for doing practical exercises*? These are the questions that guided me in selecting from the literature **rules of method** and **principles** and to dedicate one chapter to each pair**. The status of these rules and that of the principles is rather distinct and I do not expect them to be evaluated in the same way. By 'rules of method' I mean what a priori decisions should be made in order to consider all of the empirical facts provided by the specialised disciplines as being part of the domain of 'science, technology and society'. By 'principles' I mean what is *my* personal summary of the empirical facts at hand after a decade of work in this area. Thus, I expect these principles to be debated, falsified, replaced by other summaries. On the other hand, the rules of method are a package that do not seem to be easily negotiable without losing sight of the common ground I want to sketch. With them it is more a question of all or nothing, and I think they should be judged only on this ground: do they link more elements than others? Do they allow outsiders to follow science and technology further, longer and more independently? This will be the only rule of the game, that is, the only 'meta' rule that we will need to get on with our work.

---

* The present book was originally planned with exercises at the end of each chapter. For lack of space, these practical tasks will be the object of a second volume.
** Except for the first rule of method defined above. A summary of these rules and principles is given at the end of the book.

Part I

From Weaker to Stronger Rhetoric

CHAPTER 1

Literature

There are many methods for studying the fabrication of scientific facts and technical artefacts. However, the first rule of method we decided upon in the preceding Introduction is the simplest of all. We will not try to analyse the final products, a computer, a nuclear plant, a cosmological theory, the shape of a double helix, a box of contraceptive pills, a model of the economy; instead we will follow scientists and engineers at the times and at the places where they plan a nuclear plant, undo a cosmological theory, modify the structure of a hormone for contraception, or disaggregate figures used in a new model of the economy. We go from final products to production, from 'cold' stable objects to 'warmer' and unstable ones. Instead of black boxing the technical aspects of science and *then* looking for social influences and biases, we realised in the Introduction how much simpler it was to be there *before* the box closes and becomes black. With this simple method we merely have to follow the best of all guides, scientists themselves, in their efforts to close one black box and to open another. This relativist and critical stand is not imposed by us on the scientists we study; it is what the scientists themselves do, at least for the tiny part of technoscience they are working on.

To start our enquiry, we are going to begin from the simplest of all possible situations: when someone utters a statement, what happens when the others believe it or don't believe it. Starting from this most general situation, we will be gradually led to more particular settings. In this chapter, as in the following, we will follow a character, whom we will for the moment dub 'the dissenter'. In this first part of the book we will observe to what extremes a naive outsider who wishes to disbelieve a sentence is led.

Part A
Controversies

(1) Positive and negative modalities

What happens when someone disbelieves a sentence? Let me experiment with three simple cases:

(1) New Soviet missiles aimed against Minutemen silos are accurate to 100 metres.[1]

(2) Since [new Soviet missiles are accurate within 100 metres] this means that Minutemen are not safe any more, and this is the main reason why the MX weapon system is necessary.

(3) Advocates of the MX in the Pentagon cleverly leak information contending that [new Soviet missiles are accurate within 100 metres].

In statements (2) and (3) we find the same sentence (1) but inserted. We call these sentences **modalities** because they modify (or qualify) another one. The effects of the modalities in (2) and (3) are completely different. In (2) the sentence (1) is supposed to be solid enough to make the building of the MX necessary, whereas in (3) the very same statement is weakened since its validity is in question. One modality is leading us, so to speak, 'downstream' from the existence of accurate Soviet missiles to the necessity of building the MX; the other modality leads us 'upstream' from a belief in the same sentence (1) to the uncertainties of our knowledge about the accuracy of Soviet missiles. If we insist we may be led even further upstream, as in the next sentence:

(4) The undercover agent 009 in Novosibirsk whispered to the housemaid before dying that he had heard in bars that some officers thought that some of their [missiles] in ideal test conditions might [have an accuracy] somewhere between [100] and 1000 [metres] or this is at least how the report came to Washington.

In this example, statement (1) is not inserted in another phrase any more, it is broken apart and each fragment – which I have put in brackets – is brought back into a complex process of construction from which it appears to have been extracted. The directions towards which the readers of sentences (2) and (4) are invited to go are strikingly different. In the first case, they are led into the Nevada desert of the United States to look for a suitable site for the MX; in the second case they are led towards the Pentagon sifting through the CIA network of spies and disinformation. In both cases they are induced to ask different sets of questions. Following statement (1), they will ask if the MX is well designed, how much it will cost and where to locate it; believing statements (2) or (4), they will ask how the CIA is organised, why the information has been leaked, who killed agent 009, how the test conditions of missiles in Russia are set up, and so on. A reader who does not know which sentence to believe will hesitate between two attitudes; either demonstrating against the Russians for the MX or against the

CIA for a Congressional hearing on the intelligence establishment. It is clear that anyone who wishes the reader of these sentences to demonstrate against the Russians or against the CIA must make one of the statements more credible than the other.

We will call **positive modalities** those sentences that lead a statement away from its conditions of production, making it solid enough to render some other consequences necessary. We will call **negative modalities** those sentences that lead a statement in the other direction towards its conditions of production and that explain in detail why it is solid or weak instead of using it to render some other consequences more necessary.

Negative and positive modalities are in no way particular to politics. The second, and more serious, example will make this point clear:

(5) The primary structure of Growth Hormone Releasing Hormone[2] (GHRH) is Val-His-Leu-Ser-Ala-Glu-Glu-Lys-Glu-Ala.

(6) Now that Dr Schally has discovered [the primary structure of GHRH], it is possible to start clinical studies in hospital to treat certain cases of dwarfism since GHRH should trigger the Growth Hormone they lack.

(7) Dr A. Schally has claimed for several years in his New Orleans laboratory that [the structure of GHRH was Val-His-Leu-Ser-Ala-Glu-Glu-Lys-Glu-Ala]. However, by troubling coincidence this structure is also that of haemoglobin, a common component of blood and a frequent contaminant of purified brain extract if handled by incompetent investigators.

Sentence (5) is devoid of any trace of ownership, construction, time and place. It could have been known for centuries or handed down by God Himself together with the Ten Commandments. It is, as we say, a **fact**. Full stop. Like sentence (1) on the accuracy of Soviet missiles, it is inserted into other statements without further modification: no more is said about GHRH; inside this new sentence, sentence (5) becomes a closed file, an indisputable assertion, a black box. It is *because* no more has to be said about it that it can be used to lead the reader somewhere else downstream, for instance to a hospital ward, helping dwarves to grow. In sentence (7) the original fact undergoes a different transformation similar to what happened to the accuracy of Soviet missiles in statements (3) and (4). The original statement (5) is uttered by someone situated in time and space; more importantly, it is seen as something extracted from a complicated work situation, not as a gift from God but as a man-made product. The hormone is isolated out of a soup made of many ingredients; it might be that Dr Schally has mistaken a contaminant for a genuine new substance. The proof of that is the 'troubling coincidence' between the GHRH sequence and that of the beta-chain of haemoglobin. They might be homonyms, but can you imagine anybody that would confuse the order to 'release growth hormone!' with the command 'give me your carbon dioxide!'?

Depending on which sentence we believe, we, the readers, are again induced to go in opposite directions. If we follow statement (6) that takes GHRH as a fact, then we now look into possible cures for dwarfism, we explore ways of industrially producing masses of GHRH, we go into hospitals to blind-test the drug, etc. If we believe (7) we are led back into Dr Schally's laboratory in New Orleans, learning how to purify brain extracts, asking technicians if some hitch has escaped their attention, and so on. According to which direction we go, the original sentence (5) will change status: it will be either a black box or a fierce controversy; either a solid timeless certainty or one of these short-lived artefacts that appear in laboratory work. Inserted inside statement (6), (5) will provide the firm ground to do something else; but the same sentence broken down inside (7) will be one more empty claim from which nothing can be concluded.

A third example will show that these same two fundamental directions may be recognised in engineers' work as well:

(8) The only way to quickly produce efficient fuel cells[3] is to focus on the behaviour of electrodes.

(9) Since [the only way for our company to end up with efficient fuel cells is to study the behaviour of electrodes] and since this behaviour is too complicated, I propose to concentrate in our laboratory next year on the one-pore model.

(10) You have to be a metallurgist by training to believe you can tackle [fuel cells] through the [electrode] problem. There are many other ways they cannot even dream of because they don't know solid state physics. One obvious way for instance is to study electrocatalysis. If they get bogged down with their electrode, they won't move an inch.

Sentence (8) gives as a matter of fact the only research direction that will lead the company to the fuel cells, and thence to the future electric engine that, in the eyes of the company, will eventually replace most – if not all – internal combustion engines. It is then taken up by statement (9) and from it a research programme is built: that of the one-pore model. However, in sentence (10) the matter-of-fact tone of (8) is not borrowed. More exactly, it shows that (8) has not always been a matter of fact but is the result of a *decision* taken by specific people whose training in metallurgy and whose ignorance are outlined. The same sentence then proposes another line of research using another discipline and other laboratories in the same company.

It is important to understand that statement (10) does not in any way dispute that the company should get at fast and efficient fuel cells; it extracts this part of sentence (8) which it takes as a fact, and contests only the idea of studying the electrode as the best way of reaching that undisputed goal. If the reader believes in claim (9), then the belief in (8) is reinforced; the whole is taken as a package and goes where it leads the research programme, deep inside the metallurgy section of the company, looking at one-pore models of electrodes and spending years there expecting the breakthrough. If the reader believes in claim (10), then it is realised that the original sentence (8) was not *one* black box but at least *two*; the first is kept closed – fuel cells are the right goal; the other is opened – the one-pore model is an absurdity; in order to maintain the first, then the company should get into quantum physics and recruit new people. Depending on who is believed, the company may go broke or not; the consumer, in the year 2000, may drive a fuel cell electric car or not.

From these three much simpler and much less prestigious examples than the ones we saw in the Introduction, we may draw the following conclusions. A sentence may be made more of a fact or more of an artefact depending on how it is inserted into other sentences. *By itself a given sentence is neither a fact nor a fiction; it is made so by others, later on.* You make it more of a fact if you insert it as a closed, obvious, firm and packaged premise leading to some other less closed, less obvious, less firm and less united consequence. The final shape of the MX is less determined in sentence (2) than is the accuracy of Soviet missiles; the cure for dwarfism is not yet as well settled in sentence (6) as is the GHRH structure; although in sentence (9) it is certain that the right path towards fuel cells is to look at electrodes, the one-pore model is less certain than this indisputable fact. As a consequence, listeners make sentences less of a fact if they take them back where they came from, to the mouths and hands of whoever made them, or more of a fact if they use it to reach another, more uncertain goal. The difference is as great as going up or down a river. Going downstream, listeners are led to a demonstration against the Russians – see (2), to clinical studies of dwarfism – see (6), to metallurgy – see (9). Upstream, they are directed to probe the CIA – see (3), to do research in Dr Schally's laboratory – see (7), or to investigations on what quantum physics can tell us about fuel cells – see (10).

We understand now why looking at earlier stages in the construction of facts and machines is more rewarding than remaining with the final stages. Depending on the type of modalities, people will be *made to go* along completely different paths. If we imagine someone who has listened to claims (2), (6) and (9), and believed them, his behaviour would have been the following: he would have voted for pro-MX congressmen, bought shares in GHRH-producing companies, and recruited metallurgists. The listener who believed claims (3), (4), (7) and (10) would have studied the CIA, contested the purification of brain extracts, and would have recruited quantum physicists. Considering such vastly different outcomes, we can easily guess that it is around modalities that we will find the fiercest disputes since this is where the behaviour of other people will be shaped.

There are two added bonuses for us in following the earlier periods of fact construction. First, scientists, engineers and politicians constantly offer us rich material by transforming one another's statements in the direction of fact or of fiction. They break the ground for our analysis. We, laymen, outsiders and citizens, would be unable to discuss sentences (1) on the accuracy of Soviet missiles, (5) on the amino acid structure of growth hormone releasing factor, and (8) on the right way of making fuel cells. But since others dispute them and push them back into their conditions of production, we are effortlessly led to the processes of work that extract information from spies, brain soup or electrodes – processes of work we would never have suspected before. Secondly, in the heat of the controversy, specialists may themselves explain why their opponents think otherwise: sentence (3) claims that the MX partisans are *interested* in believing the accuracy of Soviet missiles; in sentence (10) the belief of the others in one absurd research project is imputed to their training as metallurgists. In other words, when we approach a controversy more closely, half of the job of interpreting the reasons behind the beliefs is already done!

(2) The collective fate of fact-making

If the two directions I outlined were so clearly visible to the eyes of someone approaching the construction of facts, there would be a quick end to most debates. The problem is that we are never confronted with such clear intersections. The three examples I chose have been arbitrarily interrupted to reveal only two neatly distinct paths. If you let the tape go on a bit longer the plot thickens and the interpretation becomes much more complicated.

Sentences (3) and (4) denied the reports about the accuracy of the Soviet missiles. But (4) did so by using a police story that exposed the inner workings of the CIA. A reply to this exposition can easily be imagined:

> (11) The CIA's certainty concerning the 100-metre accuracy of Russian missiles is not based on the agent 009's report, but on five independent sources. Let me suggest that only groups subsidised by Soviets could have an interest in casting doubts on this incontrovertible fact.

Now the readers are not sure any more where they should go from here. If sentence (4), denying the truth of sentence (1), is itself denied by (11), what should they do? Should they protest against the disinformation specialists paid by the KGB who forged sentence (4) and go on with the MX project with still more determination? Should they, on the contrary, protest against the disinformation specialists paid by the CIA who concocted (11), and continue their hearings on the intelligence gathering network with more determination? In both cases, the determination increases, but so does the uncertainty! Very quickly, the controversy becomes as complex as the arms race: missiles (arguments) are opposed by anti-ballistic missiles (counter-arguments) which are in turn counter-attacked by other, smarter weapons (arguments).

If we now turn to the second example, it is very easy to go on after sentence (7), which criticised Dr Schally's handling of GHRH, and retort:

> (12) If there is a 'troubling coincidence', it is in the fact that criticisms against Schally's discovery of GHRH are again levelled by his old foe, Dr Guillemin ... As to the homonymy of structure between haemoglobin and GHRH, so what? It does not prove Schally mistook a contaminant for a genuine hormone, no more than 'he had a fit' may be taken for 'he was fit'.

Reading (6), that assumed the existence of GHRH, you, the reader, might have decided to invest money in pharmaceutical companies; when learning of (7), you would have cancelled all plans and might have started investigations on how the Veterans Administration could support such inferior work with public funds.

But after reading the counter claims in (12), what do you do? To make up your mind you should now assess Dr Guillemin's personality. Is he a man wicked enough to cast doubt on a competitor's discovery out of sheer jealousy? If you believe so, then (7) is cancelled, which frees the original sentence (5) from doubts. If, on the contrary, you believe in Guillemin's honesty, then it is sentence (12) which is in jeopardy, and then the original claim (5) is again in danger . . ..

In this example the only thing that stands firm is this point about homonymy. At this point, to make up your mind you have to dig much further into physiology: is it possible for the blood to carry two homonymous messages to the cells without wreaking havoc in the body?

Asking these two questions – about Guillemin's integrity and about a principle of physiology – you might hear the retort (to the retort of the retort):

> (13) Impossible! It cannot be an homonymy. It is just a plain mistake made by Schally. Anyway, Guillemin has always been more credible than him. I wouldn't trust this GHRH an inch, even if it is already manufactured, advertised in medical journals, and even sold to physicians!

With such a sentence the reader is now watching a game of billiards: if (13) is true, then (12) was badly wrong, with the consequence that (7), that disputed the very existence of Schally's substance, was right, which means that (5) – the original claim – is disallowed. Naturally, the question would now be to assess the credibility of sentence (13) above. If it is uttered by an uncritical admirer of Guillemin or by someone who knows nothing of physiology, then (12) might turn out to be quite credible, which would knock (7) off the table and would thus establish (5) as an ascertained fact!

To spare the reader's patience I will stop the story here, but it is now obvious that the debate could go on. The first important lesson, here, is this: were the debate to continue, we would delve further into physiology, further into Schally's and Guillemin's personalities, and much further into the details through which hormone structures are obtained. The number of new conditions of production to tackle will take us further and further from dwarves and hospital wards. The second lesson is that with every new retort added to the debate, the status of the original discovery made by Schally in claim (5) *will be modified.* Inserted in (6) it becomes more of a fact; less when it is dislocated in (7); more with (12) that destroys (7); less again with (13); and so on. The fate of the statement, that is the decision about whether it is a fact or a fiction, depends on a sequence of debates later on. The same thing happens not only for (5), which I artificially chose as the origin of the debate, but also with each of the other sentences that qualifies or modifies it. For instance (7), which disputed Schally's ability, is itself made more of a fact with (13) that established Guillemin's honesty, but less with (12) that doubted his judgment. These two lessons are so important that this book is simply, I could argue, a development of this essential point: *the status of a statement depends on later statements.* It is made more of a certainty or less of a certainty depending on the next sentence that takes it up; this retrospective attribution is repeated for this next new sentence, which in turn might be made more of a fact or more of a fiction by a third, and so on . . .

The same essential phenomenon is visible in the third example. Before a machine is built many debates take place to determine its shape, function, or cost. The debate about the fuel cells may be easily rekindled. Sentence (10) was disputing that the right avenue to fuel cells was the one-pore electrode mode, but not that fuel cells were the right path towards the future of electric cars. A retort may come:

(14) And why get into quantum mechanics anyway? To spend millions helping physicists with their pet projects? That's bootlegging, not technological innovation, that's what it is. The electric automobile's only future is all very simple: batteries; they are reliable, cheap and already there. The only problem is weight, but if research were done into that instead of into physics, they would be lighter pretty soon.

A new pathway is proposed to the company. Physics, which for sentence (10) was the path to the breakthrough, is now the architypical dead end. The future of fuel cells, which in statements (8), (9) and (10) were packaged together with the electric car in one black box, now lies open to doubt. Fuel cells are replaced by batteries. But in sentence (14) electric cars are still accepted as an undisputable premise. This position is denied by the next claim:

(15) Listen, people will always use internal combustion engines, no matter what the cost of petrol. And you know why? Because it has got go. Electric cars are sluggish; people will never buy them. They prefer vigorous acceleration to everything else.

Suppose that you have a place on the company board that has to decide whether or not to invest in fuel cells. You would be rather puzzled by now. When you believed (9) you were ready to invest in the one-pore electrode model as it was convincingly defined by metallurgists. Then you shifted your loyalties when listening to (10) that criticised metallurgists and wished to invest in quantum physics, recruiting new physicists. But after listening to (14), you decided to buy shares in companies manufacturing traditional batteries. After listening to (15), though, if you believe it, you would be better not selling any of your General Motors shares. Who is right? Whom should you believe? The answer to this question is not in any one of the statements, but in what everyone is going to do with them later on. If you wish to buy a car, will you be stopped by the high price of petrol? Will you shift to electric cars, more sluggish but cheaper? If you do so, then sentence (15) is wrong, and (8), (9) or (10) was right, since they all wanted electric cars. If the consumer buys an internal combustion engine car without any hesitation and doubts, then claim (15) is right and all the others were wrong to invest millions in useless technologies without a future.

This retrospective transformation of the truth value of earlier sentences does not happen only when the average consumer at the end of the line gets into the picture, but also when the Board of Directors decides on a research strategy. Suppose that you 'bought the argument' presented in statement (10). You go for electric cars, you believe in fuel cells, and in quantum physics as the only way to get at them. All the other statements are *made more wrong* by this decision. The linkages between the future of the automobile, the electric engine, the fuel cells, and electrophysics are all conflated in one single black box which no one in the company is going to dispute. Everyone in the company will start from there: 'Since sentence (10) is right then let's invest so many millions.' As we will see in Chapter 3, this does not mean that your company will win. It means that, as far as you could, you shaped the other machines and facts of the past so as to win: the internal combustion engine is weakened by your decision and made more of an obsolete technology; by the same token electrophysics is strengthened, while the metallurgy section of the company is gently excluded from the picture. Fuel cells now have one more powerful ally: the Board of Directors.

Again I interrupt the controversy abruptly for practical reasons; the company may go broke, become the IBM of the twenty-first century or linger for years in limbo. The point of the three examples is that *the fate of what we say and make is in later users' hands.* Buying a machine without question or believing a fact without question has the same consequence: it strengthens the case of whatever is bought or believed, it makes it more of a black box. To disbelieve or, so to speak, 'dis-buy' either a machine or a fact is to weaken its case, interrupt its spread, transform it into a dead end, reopen the black box, break it apart and reallocate its components elsewhere. By themselves, a statement, a piece of machinery, a process are lost. By looking only at them and at their internal properties, you cannot decide if they are true or false, efficient or wasteful, costly or cheap, strong or frail. These characteristics are only gained through *incorporation* into other statements, processes and pieces of machinery. These incorporations are decided by each of us, constantly. Confronted with a black box, we take a series of decisions. Do we take it up? Do we reject it? Do we reopen it? Do we let it drop through lack of interest? Do we make it more solid by grasping it without any further discussion? Do we transform it beyond recognition? This is what happens to others' statements, in our hands, and what happens to *our* statements in others' hands. To sum up, the construction of facts and machines is a *collective* process. (This is the statement I expect *you* to believe; its fate is in your hands like that of any other statements.) This is so essential for the continuation of our travel through technoscience* that I will call it our **first principle**: the remainder of this book will more than justify this rather portentous name.

---

*In order to avoid endless 'science and technology' I forged this word, which will be fully defined in Chapter 4 only.

Part B
When controversies flare up
the literature becomes technical

When we approach the places where facts and machines are made, we get into the midst of controversies. The closer we are, the more controversial they become. When we go from 'daily life' to scientific activity, from the man in the street to the men in the laboratory, from politics to expert opinion, we do not go from noise to quiet, from passion to reason, from heat to cold. We go from controversies to fiercer controversies. It is like reading a law book and then going to court to watch a jury wavering under the impact of contradictory evidence. Still better, it is like moving from a law book to Parliament when the law is still a bill. More noise, indeed, not less.

In the previous section I stopped the controversies before they could proliferate. In real life you cannot stop them or let them go as you wish. You have to decide whether to build the MX or not; you have to know if GHRH is worth investing in; you have to make up your mind as to the future of fuel cells. There are many ways to win over a jury, to end a controversy, to cross-examine a witness or a brain extract. **Rhetoric** is the name of the discipline that has, for millennia, studied how people are made to believe and behave and taught people how to persuade others. Rhetoric is a fascinating albeit despised discipline, but it becomes still more important when debates are so exacerbated that they become scientific and technical. Although this statement is slightly counter-intuitive, it follows from what I said above. You noticed in the three examples that the more I let the controversies go on, the more we were led into what are called 'technicalities'. This is understandable since people in disagreement open more and more black boxes and are led further and further upstream, so to speak, into the conditions that produced the statements. There is always a point in a discussion when the local resources of those involved are not enough to open or close a black box. It is necessary to fetch further resources coming from other places and times. People start using texts, files, documents, articles to force others to transform what was at first an opinion into a fact. If the discussion continues then the contenders in an *oral* dispute become the *readers* of technical texts or reports. The more they dissent, the more the literature that is read will become scientific and technical. For instance, if, after reading sentence (12), which puts the accusations against the CIA into doubt, the MX is still disputed, the dissenter will now be confronted with boxes of reports, hearings, transcripts and studies. The same thing happens if you are obstinate enough not to believe in Schally's discovery. Thousands of neuroendocrinology articles are now waiting for you. Either you give up or you read them. As for fuel cells, they have their own research library whose index lists over 30,000 items, not counting the patents. This is what you have to go through in order to disagree. Scientific or technical texts–I will use the terms interchangeably–are not written differently by different breeds of writers. When you reach them, this does not mean that you quit rhetoric for the quieter realm of pure reason. It means that rhetoric has become heated enough or is still so active that many more resources have to be brought in to keep the debates going. Let me explain this by considering the anatomy of the most important and the least studied of all rhetorical vehicles: the scientific article.

(1) Bringing friends in

When an oral dispute becomes too heated, hard-pressed dissenters will very quickly allude to what others wrote or said. Let us hear one such conversation as an example:

> (16) Mr Anybody (as if resuming an old dispute): 'Since there is a new cure for dwarfism, how can you say this?'
> Mr Somebody: 'A new cure? How do you know? You just made it up.'
> –I read it in a magazine.
> –Come on! I suppose it was in a colour supplement . . .
> –No, it was in *The Times* and the man who wrote it was not a journalist but someone with a doctorate.
> –What does that mean? He was probably some unemployed physicist who does not know the difference between RNA and DNA.
> –But he was referring to a paper published in *Nature* by the Nobel Prize winner Andrew Schally and six of his colleagues, a big study, financed by all sorts of big institutions, the National Institute of Health, the National Science Foundation, which told what the sequence of a hormone was that releases growth hormone. Doesn't that mean something?
> –Oh! You should have said so first . . . that's quite different. Yes, I guess it does.

Mr Anybody's opinion can be easily brushed aside. This is why he enlists the support of a written article published in a newspaper. That does not cut much ice with Mr Somebody. The newspaper is too general and the author, even if he calls himself 'doctor', must be some unemployed scientist to end up writing in *The Times*. The situation is suddenly reversed when Mr Anybody supports his claim with a new set of allies: a journal, *Nature*; a Nobel Prize author; six co-authors; the granting agencies. As the reader can easily image, Mr Somebody's tone of voice has been transformed. Mr Anybody is to be taken seriously since he is not alone any more: a group, so to speak, accompanies him. Mr Anybody has become Mr Manybodies!

This appeal to higher and more numerous allies is often called the **argument from authority**. It is derided by philosophers and by scientists alike because it creates a majority to impress the dissenter even though the dissenter 'might be right'. Science is seen as the opposite of the argument from authority. A few win over the many because truth is on their side. The classical form of this derision is provided by Galileo when he offers a contrast between rhetoric and real science. After having mocked the florid rhetoric of the past, Galileo opposed it to what happens in physics[4]:

> But in the physical sciences when conclusions are sure and necessary and have nothing to do with human preference, one must take care not to place oneself in the defence of error; for here, a thousand Demosthenes and a thousand Aristotles would be left in the lurch by any average man who happened to hit on the truth for himself.

This argument appears so obvious at first that it seems there is nothing to add. However, a careful look at the sentence reveals two completely different arguments mixed together. Here again the two faces of Janus we have encountered in the introduction should not be confused even when they speak at once. One mouth says: 'science is truth that authority shall not overcome'; the other asks: 'how can you be stronger than one thousand politicians and one thousand philosophers?' On the left side rhetoric is opposed to science just as authority is opposed to reason; but on the right, science is a rhetoric powerful enough, if we make the count, to allow one man to win over 2000 prestigious authorities!

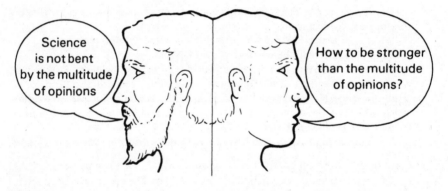

Figure 1.1

'Authority', 'prestige', 'status' are too vague to account for why Schally's article in *Nature* is stronger than Dr Nobody's piece in *The Times*. In practice, what makes Mr Somebody change his mind is exactly the opposite of Galileo's argument. To doubt that there is a cure for dwarfism, he at first has to resist his friend's opinion plus a fake doctor's opinion plus a newspaper. It is easy. But at the end, how many people does he have to oppose? Let us count: Schally and his coworkers plus the board of the New Orleans university who gave Schally a professorship plus the Nobel Committee who rewarded his work with the highest prize plus the many people who secretly advised the Committee plus the editorial board of *Nature* and the referees who chose this article plus the scientific boards of the National Science Foundation and of the National Institutes of Health who awarded grants for the research plus the many technicians and helping hands thanked in the acknowledgements. That's a lot of people and all this is *before* reading the article, just by counting how many people are engaged in its publication. For Mr Somebody, doubting Mr Anybody's opinion takes no more than a shrug of the shoulders. But how can you shrug off dozens of people whose honesty, good judgment and hard work you must weaken before disputing the claim?

The adjective 'scientific' is not attributed to *isolated* texts that are able to oppose the opinion of the multitude by virtue of some mysterious faculty. A document becomes scientific when its claims stop being isolated and when the number of people engaged in publishing it are many and explicitly indicated in the text. When reading it, it is on the contrary the reader who becomes *isolated*. The careful marking of the allies' presence is the first sign that the controversy is now heated enough to generate technical documents.

(2) Referring to former texts

There is a point in oral discussions when invoking other texts is not enough to make the opponent change his or her mind. The text itself should be brought in and read. The number of external friends the text comes with is a good indication of its strength, but there is a surer sign: references to other documents. The presence or the absence of references, quotations and footnotes is so much a sign that a document is serious or not that you can transform a fact into fiction or a fiction into fact just by adding or subtracting references.

The effect of references on persuasion is not limited to that of 'prestige' or 'bluff'. Again, it is a question of *numbers*. A paper that does not have references is like a child without an escort walking at night in a big city it does not know: isolated, lost, anything may happen to it. On the contrary, attacking a paper heavy with footnotes means that the dissenter has to weaken each of the other papers, or will at least be threatened with having to do so, whereas attacking a naked paper means that the reader and the author are of the same weight: face to face. The difference at this point between technical and non-technical literature is not that one is about fact and the other about fiction, but that the latter gathers only a few resources at hand, and the former a lot of resources, even from far away in time and space. Figure 1.2 drew the references reinforcing another paper by Schally.[5]

Whatever the text says we can see that it is already linked to the contents of no less than thirty-five papers, from sixteen journals and books from 1948 to 1971. If you wish to do anything to this text and if there is no other way of getting rid of the argument you know in advance that you might have to engage with all these papers and go back in time as many years as necessary.

However, stacking masses of reference is not enough to become strong if you are confronted with a bold opponent. On the contrary, it might be a source of weakness. If you explicitly point out the papers you attach yourself to, it is then possible for the reader – if there still are any readers – to trace each reference and to probe its degree of attachment to your claim. And if the reader is courageous enough, the result may be disastrous for the author. First, many references may

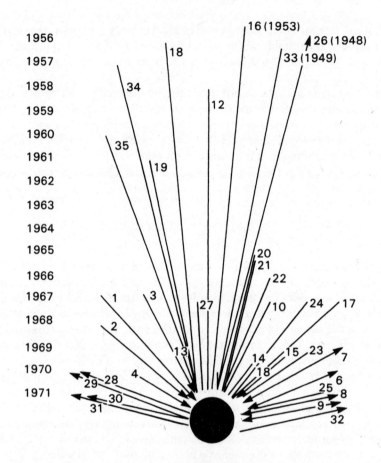

Schally's article

—those going to the text are constituting the imported paradigm;
—those going from the text are discussing the referred papers (only one, 32, is critical)
—those going both ways refer to previous work by the same group on the same question

Figure 1.2

be misquoted or wrong; second, many of the articles alluded to might have no bearing whatsoever on the claim and might be there just for display; third, other citations might be present but only because they are always present in the author's articles, whatever his claim, to mark affiliation and show with which group of scientists he identifies – these citations are called **perfunctory**.[6] All these little defects are much less threatening for the author's claim than the references to papers which explicitly say the contrary of the author's thesis. For instance, Figure 1.2 shows Schally referring to the following paper (reference number 32):

(17) 32. Veber, D.F., Bennett, C., Milkowski, J.D., Gal, G., Denkewalter, R.D., and Hirschman, R., in *Biochemistry and Biophysics Communication*, 45, 235 (1971).

This is a quite an impressive set of allies, *if* they support the claim. But the author should not let the unflinching reader go to reference 32 by himself. Why not? Because in this paper Veber *et al.* link the structure of Schally's GHRH with that of the beta-chain of haemoglobin, levelling exactly the criticisms that we have already seen in sentence (7). A dangerous link indeed in an opponent's hands. To ward it off, Schally cites it but qualifies the paper within his own text:

> (18) [Note added in proof.] D.F. Veber et al. have pointed out the similarity between the structure of our decapeptide and the amino-terminal of the Beta-chain of porcine haemoglobin (ref. 32). The significance of this observation remains to be established.

The article is not only referred to; it is also qualified or, as we said earlier, modalised. In this case, the reader is warned not to take Veber's article as a fact; since its significance is not established, it cannot be used against Schally to destroy his GHRH (remember that if Veber's claims were turned into a fact, then Schally's own article would become just a fiction). What Schally does to sentence (17) is done by all articles to all their references. Instead of passively linking their fate to other papers, the article *actively* modifies the status of these papers. Depending on their interests, they turn them more into facts or more into fictions, thus replacing crowds of uncertain allies by well-arrayed sets of obedient supporters. What is called the **context of citation** shows us how one text acts on others to make them more in keeping with its claims.

In sentence (18) Schally added the other article referred to in excerpt (17) to maintain it in a stage intermediate between fact and fiction. But he also needs well-established facts so as to start his article with a black box which no one would dare to open. This solid foundation is offered, not surprisingly, at the beginning of the article:

> (19) The hypothalamus controls the secretion of growth hormone from the anterior pituitary gland (ref. 1 to Pend Muller, E.E., *Neuroendocrinology*, 1, 537, 1967). This control is mediated by a hypothalamic substance designated growth hormone releasing hormone (ref. 2 to Schally, A.V., Arimura, A., Bowers, C.Y., Kastin, A.J., Sawano, S., and Redding, T.W., *Recent Progress in Hormone Research*, 24, 497, 1968).

The first reference is borrowed as it stands with no indication of doubt or uncertainty. Besides, it is a five-year-old citation – a very long time for these short-lived creatures. If you, the reader, doubt this control of the hypothalamus, then forget it, you are out of the game entirely. Inside neuroendocrinology, this is the most solid point, or, as it is often called, the **paradigm**.[7] The second reference is also borrowed as a matter of fact, although it is slightly weaker than the former. Dissent was impossible to reference 1, at least coming from a neuro-endocrinologist; with reference 2 it is possible for a colleague to nitpick: maybe the control is mediated by something other than a hormone; maybe, even if it is a hormone, it blocks growth hormone instead of triggering it; or, at the very least, the name Schally gave to this substance could be criticised (Guillemin, for instance, calls it GRF). No matter what controversy could start here, Schally needs this reference in his article as a fact, since without it the whole paper would be purposeless: why look for a substance if the possibility of its existence is denied? Let us not forget that, according to our first principle, by borrowing references 1 and 2 as matters of fact he makes them more certain, strengthening their case as well as his own.

There are many other papers this article needs to borrow without question, especially the ones describing methods used in determining the sequence of peptides in general. This is visible in another excerpt from the same article:

> (20) The porcine peptide used in this work was an essentially homogeneous sample isolated as described previously (refs. 5, 9). (. . .) In some cases products of carboxypeptidase B. were analysed with the lithium buffer system of Benson, Gordon and Patterson (ref. 10). (. . .) The Edman degradation was performed as reported by Gottlieb et al. (ref. 14). The method of Gray and Smith (ref. 15) was also used.

None of these references, contrary to the others, are qualified either positively or negatively. They are simply there as so many signposts indicating to the readers, if need be, the technical resources that are under Schally's command. The reader who would doubt the hormone sequence is directed towards another set of people: Benson, Edman, Gottlieb, and even Gray and Smith. The work of these people is not present in the text, but it is indicated that they could be mobilised at once if need be. They are, so to speak, in reserve, ready to bring with them the many technical supports Schally needs to make his point firm.

Although it is convenient for a text to borrow references that could help in strengthening a case, it is also necessary for a text to attack those references that could explicitly oppose its claims. In sentence (18) we saw how the referred paper was maintained in a state between fact and fiction, but it would have been better to destroy it entirely so as to clear the way for the new paper. Such a destruction happens in many ways directly or obliquely depending on the field and the authors. Here is an instructive negative modality made by Guillemin about a set of papers, including the one written by Schally that we just studied:

> (21) The now well established concept of a neurohumoral control of adenohypophyseal secretions by the hypothalamus indicates the existence of a hypothalamic growth-hormone-releasing factor (GRF) (ref. 1) having somatostatin as its inhibitory counterpart (ref. 2). So far hypothalamic GRF has not been unequivocally characterized, despite earlier claims to the contrary (ref. 3).

This citation comes from a recent paper by Guillemin, presenting a new structure for the same GHRH, which he calls GRF. Reference 3 is to Schally's paper. The beginning of excerpt (21) is the same as that of (19) in Schally's text: the hypothalamic control is the blackest of all black boxes. Even if they are in dispute with one another Schally and Guillemin accept that no one can contest this control and call him or herself a neuroendocrinologist. But Schally's article in Guillemin's hands is not a black box at all. If Schally's sequence had been a fact, then the 1982 article by Guillemin would be meaningless. It would also be meaningless if Schally's sequence had any relation with Guillemin's. The latter would just add to the former's work. With sentence (21) Guillemin's paper just pushes aside Schally's sequence. It was not an unequivocal fact, but a very equivocal 'claim'. It does not count; it was a blind alley. Real work starts from this 1982 paper, and real GRF (wrongly called by Schally GHRH) starts from this sequence.

Articles may go still further in transforming the former literature to their advantage. They might combine positive and negative modalities, strengthening for instance a paper X in order to weaken a paper Y that would otherwise oppose their claim. Here is an instance of such a tactic:

> (22) A structure has been proposed for GRF [reference to Schally's article]; it has been recently shown, however [reference to Veber *et al.*] that it was not GHRH but a minor contaminant, probably a piece of hemoglobin.

Veber's article, that Schally himself cited in excerpt (18), did not say exactly what it is *made to say* here; as for Schally's article it did not exactly claim to have found the GHRH structure. This does not matter for the author of sentence (22); he simply needs Veber as an established fact to make Schally's paper more of an empty claim which, after a rebound, gives more solidity to sentence (21) that proposes a new real substance 'despite earlier claims to the contrary'.

Another frequent tactic is to oppose two papers so that they disable one another. Two dangerous counter-claims are turned into impotent ones. Schally, in the paper under study, uses one test in order to assay his GHRH. Other writers who tried to replicate his claim had used another type of test, called the radioimmunoassay, and failed to replicate Schally's claim. That is a major problem for Schally, and in order to find a way out he retorts that:

> (23) This synthetic decapeptide material or the natural material were (sic) only weakly active in tests where the release of growth hormone was measured by a radioimmunoassay for rat growth hormone (two refs.). However, the adequacy of radioimmunoassays for measuring rat growth hormone in plasma has been questioned recently (ref. 8).

Could the absence of any effect of GHRH in the assay not shake Schally's claim? No, because another paper is used to cast doubt on the assay itself: the absence of GHRH proves nothing at all. Schally is relieved.

It would be possible to go much further in the Byzantine political schemes of the context of citations. Like a good billiard player, a clever author may calculate shots with three, four or five rebounds. Whatever the tactics, the general strategy is easy to grasp: do whatever you need to the former literature to render it as helpful as possible for the claims you are going to make. The rules are simple enough: weaken your enemies, paralyse those you cannot weaken (as was done in sentence (18)), help your allies if they are attacked, ensure safe communications with those who supply you with indisputable instruments (as in (20)), oblige your enemies to fight one another (23); if you are not sure of winning, be humble and understated. These are simple rules indeed: the rules of the oldest politics. The result of this adaptation of the literature to the needs of the text is striking for the readers. They are not only impressed by the sheer quantity of references; in addition, all of these references are aimed at specific goals and arrayed for one purpose: lending support to the claim. Readers could have resisted a crowd of disorderly citations; it is much harder to resist a paper which has carefully modified the status of all the other articles it puts to use. This activity of the scientific paper is visible in Figure 1.3 in which the paper under study is a point related by arrows to the other papers, each type of arrow symbolising a type of action in the literature.

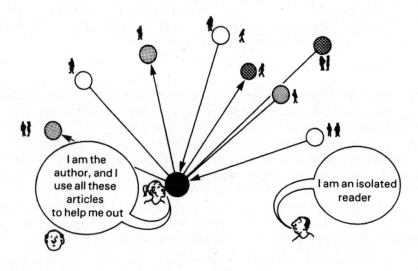

Figure 1.3

(3) Being referred to by later texts

The goal of convincing the reader is not automatically achieved, even if the writer has a high status, the references are well arrayed, and the contrary evidences are cleverly disqualified. All this work is not enough for one good reason: whatever a paper does to the former literature, the later literature will do to it. We saw earlier that a statement was fact or fiction not by itself but only by what the other sentences made of it later on. To survive or to be turned into fact, a statement needs the *next generation* of papers (I will call 'generation' the span of time necessary for another round of papers to be published that refers to the first ones, that is between two and five years). Metaphorically speaking, statements, according to the first principle, are much like genes that cannot survive if they do not manage to pass themselves on to later bodies. In the former section we saw how Schally's paper inserted other articles, distributing honour and shame, disabling some, strengthening others, borrowing without qualification from still more papers, and so on. All of the cited papers survive in Schally's paper and are modified by its action. But no paper is strong enough to stop controversies. By definition, a fact cannot be so well established that no support is necessary any more. That would be like saying that a gene is so well adapted that it does not need new bodies to survive! Schally may adapt the literature to his end; but each of his assertions, *in turn*, needs other articles later on to make it more of a fact. Schally cannot avoid this any more than the papers he quoted could survive without his taking them up.

Remember how in claim (18) Schally needed the harsh criticisms formulated in Veber's article cited in (17) to remain uncertain so as to protect his claim against a fatal blow. But to maintain (17) in such a state, Schally needs others to confirm his action. Although Schally is able to control most of what he writes in his papers, he has only weak control over what others do. Are they going to follow him?

One way to answer this is to examine the references in *other articles* subsequent to Schally's paper and to look at *their* context of citation. What did they do with what Schally did? It is possible to answer this question through a bibliometric instrument called the *Science Citation Index*.[8] For instance, statement (17) is not maintained by later articles in between fact and fiction. On the contrary, every later writer who cites it takes it as a well-established fact, and they all say that haemoglobin and GHRH have the same structure, using this fact to undermine Schally's claim to have 'discovered' GHRH (this is now placed in quotation marks). If, in the first generation, Schally was stronger than Veber – see (18) – and since there was no ally later on to maintain this strength, in the next generation it is Veber who is strong and Schally who made a blunder by taking a trivial contaminant for a long-sought-after hormone. This reversal is imposed by the other papers and the way *they in turn transform the earlier literature to suit their needs.* If we add to Figure 1.3 a third generation we obtain something like what is shown in Figure 1.4.

By adding the later papers we may map out how the actions of one paper are supported or not by other articles. The result is a cascade of transformations, each of them expecting to be confirmed later by others.

We now understand what it means when a controversy grows. If we wished to continue to study the dispute we will not have simply to read one paper alone and possibly the articles to which it refers; we will also be bound to read all the others that convert each of the operations made by the first paper towards the state of fact or that of fiction. The controversy swells. More and more papers are involved in the mêlée, each of them positioning all the others (fact, fiction, technical details), but no one being able to fix these positions *without the help of the others.* So more and more papers, enrolling more and more papers, are needed at each stage of the discussion – and the disorder increases in proportion.

There is something worse, however, than being criticised by other articles; it is being misquoted. If the context of citations is as I have described, then this misfortune must happen quite often! Since each article adapts the former literature to suit its needs, all deformations are fair. A given paper may be cited by others for completely different reasons in a manner far from its own interests. It may be cited without being read, that is perfunctorily; or to support a claim which is exactly the opposite of what its author intended; or for technical details so minute that they escaped their author's attention; or because of intentions attributed to the authors but not explicitly stated in the text; or for many other reasons. We cannot say that these deformations are unfair and that each paper should be read honestly as it is; these deformations are simply a consequence of what I called the activity of the papers on the literature; they all manage to do the same carving out of the literature to put their claims into as favourable as possible a state. If any of these operations is taken up and accepted by the others as a fact, then that's it; it is a fact and not a deformation, however much the author may protest. (Any reader who has ever written a quotable article in any discipline will understand what I mean.)

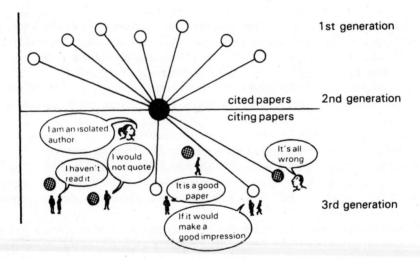

Figure 1.4

There is something still worse, however, than being either criticised or dismantled by careless readers: it is being *ignored*. Since the status of a claim depends on later users' insertions, what if there are *no* later users whatsoever? This is the point that people who never come close to the fabrication of science have the greatest difficulty in grasping. They imagine that all scientific articles are equal and arrayed in lines like soldiers, to be carefully inspected one by one. However, most papers are never read at all. No matter what a paper did to the former literature, if no one else does anything with it, then it is as if it never existed at all. You may have written a paper that settles a fierce controversy once and for all, but if readers ignore it, it cannot be turned into a fact; it simply *cannot*.

You may protest against the injustice; you may treasure the certitude of being right in your inner heart; but it will never go further than your inner heart; you will never go further in certitude without the help of others. Fact construction is so much a collective process that an isolated person builds only dreams, claims and feelings, not facts. As we will see later in Chapter 3, one of the main problems to solve is to interest someone enough to be read at all; compared to this problem, that of being believed is, so to speak, a minor task.

In the turmoil generated by more and more papers acting on more and more papers, it would be wrong to imagine that everything fluctuates. Locally, it happens that a few papers are always referred to by later articles with similar positive modalities, not only for one generation of articles but for several. This event – extremely rare by all standards – is visible every time a claim made by one article is borrowed without any qualification by many others. This means that anything it did to the former literature is turned into fact by whoever borrows it later on. The discussion, at least on this point, is ended. A black box has been produced. This is the case of the sentence 'fuel cells are the future of electric cars' inserted inside statements (8), (9) and (10). It is also the case for the control by the hypothalamus of growth hormone. Although Schally and Guillemin disagree on many things, this claim is borrowed by both without any qualification or misgivings – see sentences (19) and (20). In Figure 1.5 illustrating the context of

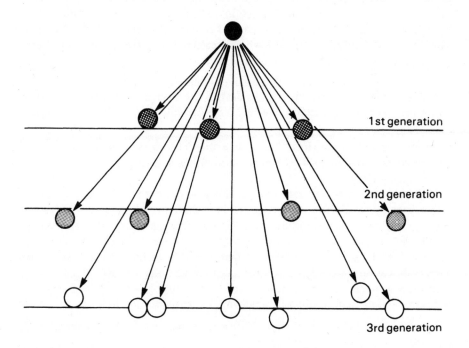

Figure 1.5

citations such an event will be seen as a regular flow of arrows all aligned in the same direction and leading to more and more papers. Every new paper getting into the fray pushes it one step further, adding its little force to the force of the already established fact, rather than reversing the trend.

This rare event is what people usually have in mind when they talk of a 'fact'. I hope it is clear by now that this event does not make it qualitatively different from fiction; a fact is what is collectively stabilised from the midst of controversies when the activity of later papers does not consist only of criticism or deformation but also of confirmation. The strength of the original statement does not lie in itself, but is derived from any of the papers that incorporate it. In principle, any of the papers could reject it. The control of growth hormone by the hypothalamus could be disputed, it has been, it will be disputed; but to do so the dissenter will be faced not with one claim in one paper, but with the same claims incorporated in hundreds of papers. It is not impossible in principle; it is just enormously difficult in practice. Each claim comes to the future author with its history, that is with itself plus all the papers that did something with it or to it.

This activity of each of the papers that makes up the strength of a given article is made visible not by any criticism – since in this case there is none – but by the erosion the original statement submits to. Even in the very rare cases where a statement is continuously believed by many later texts and borrowed as a matter of fact, it does not stay the same. The more people believe it and use it as a black box the more it undergoes transformations. The first of these transformations is an extreme *stylisation*. There is a mass of literature on the control of growth hormone, and Guillemin's article which I referred to is five pages long. Later papers, taking his article as a fact, turn it into one sentence:

> (24) Guillemin et al. (ref.) have determined the sequence of GRF: H Tyr Ala Asp Ala Ile Phe Thr Asn Ser Tyr Arg Lys Val Leu Gly Gln Leu Ser Ala Arg Lys Leu Leu Gln Asp Ile Met Ser Arg Gln Gln Gly Gly Ser Asn Gln Glu Arg Gly Ala Arg Ala Arg Leu NH2.

Later on this sentence itself is turned into a one-line long statement with only one simplified positive modality: 'X (the author) has shown that Y.' There is no longer any dispute.

If sentence (24) is to continue to be believed, as opposed to (5), each successive paper is going to add to this stylisation. The activity of all the later papers will result in the name of the author soon being dropped, and only the reference to Guillemin's paper will mark the origin of the sequence. This sequence in turn is still too long to write. If it becomes a fact, it will be included in so many other papers that soon it would not be necessary to write it at all or even to cite such a well-known paper. After a few dozen papers using statement (24) as an incontrovertible fact, it will be transformed into something like:

> (25) We injected sixty 20-day-old Swiss albino male mice with synthetic GRF . . . etc.

The accepted statement is, so to speak, eroded and polished by those who accept it. We are back to the single sentence statements with which I started this chapter – see (1), (5) and (8). Retrospectively, we realise that a lot of work went into this stylisation and that a one-phrase fact is never at the beginning of the process (as I had to imply in order to get our discussion going) but is already a semi-final product. Soon, however, the reference itself will become redundant. Who refers to Lavoisier's paper when writing the formula H_2O for water? If positive modalities continue acting on the same sentence (24), then it will become so well known that it will not be necessary even to talk about it. The original discovery will have become *tacit knowledge*. GRF will be one of the many vials of chemicals that any first year university student takes from the shelf at some point in his or her training. This erosion and stylisation happens only when all goes well; each successive paper takes the original sentence as a fact and encapsulates it, thereby pushing it, so to speak, one step further. The opposite happens, as we saw earlier, when negative modalities proliferate. Schally's sentence (5) about a new GHRH was not stylised and was still less incorporated into tacit practice. On the contrary, more and more elements he would have liked to maintain as tacit emerge and are talked about, like the purification procedures of statement (7) or his previous failures in (13). Thus, depending on whether the other articles push a given statement downstream or upstream, it will be incorporated into tacit knowledge with no mark of its having been produced by anyone, or it will be opened up and many specific conditions of production will be added. This double move with which we are now familiar is summarised in Figure 1.6 and allows us to take our bearings in any controversy depending on which stage the statement we chose as our point of departure happens to be and in which direction other scientists are pushing it.

Now we start to understand the kind of world into which the reader of scientific or technical literature is gradually led. Doubting the accuracy of Soviet missiles, (1), or Schally's discovery of GHRH, (5), or the best way to build fuel cells, (8), was at first an easy task. However, if the controversy lasts, more and more elements are brought in, and it is no longer a simple verbal challenge. We go from conversation between a few people to texts that soon fortify themselves, fending off opposition by enrolling many other allies. Each of these allies itself uses many different tactics on many other texts enlisted in the dispute. If no one takes up a paper, it is lost forever, no matter what it did and what it cost. If an article claims to finish the dispute once and for all it might be immediately dismembered, quoted for completely different reasons, *adding* one more empty claim to the turmoil. In the meantime, hundreds of abstracts, reports and posters get into the fray, adding to the confusion, while long review papers strive to put some order into the debates though often on the contrary simply adding more fuel to the fire. Sometimes a few stable statements are borrowed over and over again by many papers but even in these rare cases, the statement is slowly eroded, losing its original shape, encapsulated into more and more foreign statements, becoming so familiar and routinised that it becomes part of tacit practice and disappears from view!

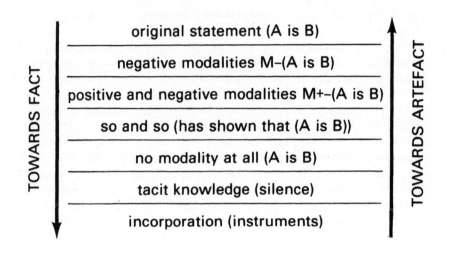

Figure 1.6

This is the world with which someone who wishes to dissent and make a contribution to the debates will be confronted. The paper he or she is reading has braced itself for survival in this world. What must it do in order to be read, to be believed, to avoid being misunderstood, destroyed, dismembered, ignored? How can it ensure that it is taken up by others, incorporated into later statements as a matter of fact, quoted, remembered and acknowledged? This is what has to be sought by the authors of a new technical paper. They have been led by the heated controversy into reading more and more articles. Now they have to *write* a new one in order to put to rest whichever issue they started from: the MX affair, the GHRH blunder, the fuel cell fiasco. Needless to say that, by now, *most dissenters will have given up*. Bringing friends in, launching many references, acting on all these quoted articles, visibly deploying this battlefield, is already enough to intimidate or to force most people out. For instance, if we wish to dispute the accuracy of Soviet missiles as in (1), the discovery of GHRH as in (5) or the right way to get at fuel cells as in (8), we will be very, very isolated. I do not say that because the literature *is too technical* it puts people off, but that, on the contrary, we feel it necessary to call technical or scientific a literature that is made to isolate the reader by bringing in many more resources. The 'average man who happens to hit the truth', naively postulated by Galileo, will have no chance to win over the thousands of articles, referees, supporters and granting bodies who oppose his claim. The power of rhetoric lies in making the dissenter feel lonely. This is indeed what happens to the 'average man' (or woman) reading the masses of reports on the controversies we so innocently started from.

Part C
Writing texts that withstand
the assaults of a hostile environment

Although most people will have been driven away by the external allies invoked by the texts, Galileo is still right, because a few people may not be willing to give up. They may stick to their position and not be impressed by the title of the journal, the names of authors, or by the number of references. They will read the articles and still dispute them. The image of the scientific David fighting against the rhetorical Goliath reappears and gives some credence to Galileo's position. No matter how impressive the allies of a scientific text are, this is not enough to convince. Something else is needed. To find this something else, let us continue our anatomy of scientific papers.

(1) Articles fortify themselves

For a few obstinate readers, already published articles are not enough: more elements have to be brought in. The mobilisation of these new elements transforms deeply the manner in which texts are written: they become more technical and, to make a metaphor, **stratified**. In sentence (21), I quoted the beginning of a paper written by Guillemin. First, this sentence mobilised a two-decade-old fact, the control by the hypothalamus of the release of growth hormone, and then a decade-old fact, the existence of a substance, somatostatin, that inhibits the release of growth hormone. In addition, Schally's claim about this new substance was dismissed. But this is not enough to make us believe that Guillemin has done better than Schally and that his claim should be taken more seriously than that of Schally. If the beginning of his paper was playing on the existing literature in the manner I analysed above, it soon becomes very different. The text announces, for instance, more material from which to extract these elusive substances. The authors found a patient with enormous tumours formed in the course of a rare disease, acromegaly, these tumours producing large quantities of the sought-for substance.[9]

> (26) At surgery, two separate tumors were found in the pancreas (ref. 6); the tumor tissues were diced and collected in liquid nitrogen within 2 or 5 minutes of resection with the intent to extract them for GRF. (. . .) The extract of both tumors contained growth hormone releasing activity with the same elution volume as that of hypothalamic GRF (K_{av}=0,43, where K_{av} is the elution on constant (ref. 8). The amounts of GRF activity (ref. 9) were minute in one of the tumors (0.06 GRF unit per milligram (net weight), but extremely high in the other (1500 GRF units per milligram (net weight), 5000 times more than we had found in rat hypothalamus (ref. 8).

Now, we are in business! Sentence (26) appears to be the most difficult sentence we have had to analyse so far. Where does the difficulty come from? From the number of objections the authors have to prevent. Reading it after the other sentences, we have not suddenly moved from opinions and disputes to facts and technical details; we have reached a state where the discussion is so tense that each word fences off a possible fatal blow. Going from the other disputes to this one is like going from the first elimination rounds to the final match at Wimbledon. Each word is a move that requires a long commentary, not because it is 'technical', but because it is the final match *after* so many contests. To understand this, we simply have to add the reader's objection to the sentence that answers it. This addition transforms sentence (26) into the following dialogue:

(27)– How could you do better than Schally with such minute amounts of your substance in the hypothamali?
– We find tumours producing masses of substance making isolation much easier than anything Schally could do.
– Are you kidding? These are pancreas tumours, and you are looking for a hypothalamic substance that is supposed to come from the brain!
– Many references indicate that often substances from the hypothalamus are found in the pancreas too, but anyway they have the same elution volume; this is not decisive but it is quite a good proof – enough, at any rate, to accept the tumour as it is, with an activity 5000 times greater than hypothalamic. No one can deny that it is a godsend.
– Hold on! How can you be so sure of this 5000; you cannot just conjure up figures? Is it dry weight or wet weight? Where does the standard come from?
– Okay. First, it is dry weight. Second, one GRF unit is the amount of a purified GRF preparation of rat hypothalamic origin that produces a half-maximal stimulation of growth hormone in the pituitary cell monolayer bioassay. Are you satisfied?
– Maybe, but how can we be sure that these tumours have not deteriorated after the surgery?
– We told you, they were diced and put in liquid nitrogen after 2 to 5 minutes. Where could you find better protection?

Reading the sentences of the paper without imagining the reader's objections is like watching only one player's strokes in the tennis final. They just appear as so many empty gestures. The accumulation of what appears as technical detail is not meaningless; it is just that it makes the opponent harder to beat. The author protects his or her text against the reader's strength. A scientific article becomes more difficult to read, just as a fortress is shielded and buttressed; not for fun, but to avoid being sacked.

Another deep transformation occurs in the texts that want to be strong enough to resist dissent. So far, the sentences we studied linked themselves to *absent* articles or events. Every time the opponent started to doubt, he or she was sent back to other texts, the link being established either by the references or sometimes by quotations. There is, however, a much more powerful ploy, and it is to *present* the very thing you want the readers to believe in the text. For instance:

(28) Final purification of this material by analytical reverse-phase HPLC yielded three highly purified peptides with GRF activity (Fig. 1)

The authors are not asking you to believe them. They do not send you back outside the texts to libraries to do your homework by reading stacks of references, but to figure 1 within the article:

(29)

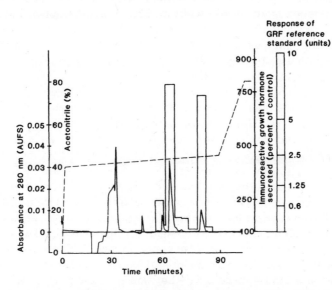

This figure *shows* what the text *says*, but is not quite transparent for all readers, even for the few who are left in the controversy. Then another text, the legend, explains how to read the figure, as the name 'legend' indicates:

(30) Final purification of hpGRF by reverse-phase HPLC. The column (Ultrasphere C18), 25 by 0,4cm, 5-(pu)m particle size, was eluted with a gradient of acetonitrile (---) in 0.5 percent (by volume) heptafluorobutyric acid at a flow-rate of 0.6ml/min. Fractions (2.4 ml) were collected as indicated on the abscissa and portions were used for bioassays (ref. 7). The vertical bars represent the amount of growth hormone secreted in the assay of each fraction of the effluent, expressed as percentage of the amount of growth hormone secreted by the pituitary cells receiving no treatment. AUFS, absorbance units full scale.

The reader was sent from statement (28) to excerpt (29) and from there to the legend (30). The text said that 'three purified peptides had GRF activity'; what is seen in figure 1 is the superimposition of peaks and vertical bars. 'Peaks' and 'bars' are said in the legend to be the visual equivalents of 'purity' and 'activity'. Belief in the author's *word* is replaced by the inspection of 'figures'. If there is any doubt about where the picture comes from, then sentence (30), the legend, will offer a new line of support. Peaks are not a visual display chosen by chance; they are what is drawn by an instrument (called a High Pressure Liquid Chromatograph); if the reader knows anything about the instrument and how different pictures can be obtained from it, then details are provided to hold the image steady: the size of particles, the timing, the conventions for drawing the lines, and so on.

What is gained in persuasion, by arraying excerpts (28), (29) and (30) in tiers? The dissenter is now faced not only with the author's opinion, not only with older articles' positions, but also with what the text is about. Often, when we talk, we designate absent things, which we call the **referent** of our speech. 'Six peach trees blooming' is a phrase about trees which I am not showing you. The situation is completely different when sentence (28) claims that three active and pure substances exist. The referent of this sentence is immediately added to the commentary; it is the figure shown in (29), and so is the referent of this referent, the legend (30). This transformation of the usual literature is a sure indicator that we are now faced with a technical or a scientific text. In this kind of literature you may, so to speak, have your cake and eat it too. The effects on conviction are enormous. The assertion 'we discovered GRF' does not stand *by itself*. It is supported first by many other texts and second by the author's assertions. This is good, but not enough. It is much more powerful if the supporters are arrayed in the text itself. How can you deny statement (28)? Look for yourself at the peaks in (29)! You are doubtful about the meaning of the figure? Well, read the legend. You only have to believe the evidence of your own eyes; this is not a question any more of belief; this is *seeing*. Even doubting Thomas would abandon his doubts (even though you cannot touch GRF – but wait until the next chapter . . .).

We are certain now that the texts we have been led to by the intensity of the controversies are scientific. So far, journalists, diplomats, reporters and lawyers could have written texts with references and with careful labelling of the authors' roles, titles and sources of support. Here, we enter another game entirely. Not because the prose is suddenly written by extraterrestrial minds, but because it tries to pack inside the text as many supporters as possible. This is why what is often called 'technical details' proliferate. The difference between a regular text in prose and a technical document is the stratification of the latter. The text is arranged in layers. Each claim is interrupted by references outside the texts or inside the texts to other parts, to figures, to columns, tables, legends, graphs. Each of these in turn may send you back to other parts of the same texts or to more outside references. In such a stratified text, the reader, once interested in reading it, is as free as a rat in a maze.

The transformation of linear prose into, so to speak, a folded array of successive defence lines is the surest sign that a text has become scientific. I said that a text without references was naked and vulnerable, but even with them it is weak as long as it is not stratified. The simplest way to demonstrate this change in solidity is to look at two articles in the same field taken at a twenty-year interval. Compare for instance the first primatology articles written by the pioneers of this field twenty years ago with one recent application of sociobiology to the study of primates written by Packer.[10] Visually, and even without reading the article, the difference is striking. In both cases, it is about baboons, but the prose of the first article flows with no interruption except sparse references and a few pictures of baboons (like the ones you could find in a journalist's travel account); Packer's article, on the contrary, is stratified into many layers. Each observation of baboons is coded, sifted for its statistical significance; curves and diagrams summarise columns; no part of the paper stands by itself but each is linked by many references to other layers (Methods, Results, Discussion). Comparing Hall's and Packer's texts is like comparing a musket with a machine-gun. Just by looking at the differences in prose you can imagine the sort of worlds they had to write in: Hall was alone, one of the first baboon watchers; Packer is in a pack of scientists who watch closely not only baboons but also one another! His prose folds itself into many defensive layers to withstand their objections.

Notice that neither in Packer's nor in Guillemin's and Schally's articles do you see the actual furry creatures called 'baboons' or the 'GHRH'. Nevertheless, through their stratification, these articles give the reader an impression of *depth of vision*; so many layers supporting each other create a thicket, something you cannot breach without strenuous efforts. This impression is present even when the text is later turned into an artefact by colleagues. No one getting into the GRF business or into baboon study can now write in plain naked prose, no matter what he or she sees and wants. It would be like fighting tanks with swords. Even people who wish to defraud have to pay an enormous price in order to create this depth that resembles reality. Spector, a young biologist convicted of having fudged his data, had to hide his fraud in a four-page long section on Materials and Methods.[11] Inside the array of hundreds of methodological precautions only one sentence is fabricated. It is, so to speak, a homage rendered by vice to virtue, since such a fraud is not within the reach of just any crook!

At the beginning of this section, we said that we needed 'something other' than just references and authorities to win over the dissident. We understand now that going from the outer layers of the articles to the inner parts is not going from the argument of authority to Nature as it is going from authorities to more authorities, from numbers of allies and resources to still *greater numbers*. Someone who disbelieves Guillemin's discovery will now be faced not only with big names and thick references, but also with 'GRF units', 'elution volume', 'peaks and bars', 'reverse-phase HPLC'. Disbelieving will not only mean courageously fighting masses of references, but also unravelling endless new links that tie instruments, figures and texts together. Even worse, the dissenter will be unable to oppose the text to the real world out there, since the text claims to bring within it the real world 'in there'. The dissenter will indeed be isolated and lonely since the referent itself has passed into the author's camp. Could it hope to break the alliances between all these new resources inside the article? No, because of the folded, convoluted and stratified form the text has taken defensively, tying all its parts together. If one doubts figure 1 in excerpt (29), then one has to doubt reverse phase HPLC. Who wishes to do so? Of course, any link can be untied, any instrument doubted, any black box reopened, any figure dismissed, but the accumulation of allies in the author's camp is quite formidable. Dissenters are human too; there is a point where they cannot cope against such high odds.

In my anatomy of scientific rhetoric I keep shifting from the isolated reader confronted by a technical document to the isolated author launching his document amidst a swarm of dissenting or indifferent readers. This is because the situation is symmetrical: if isolated, the author should find new resources to convince readers; if he or she succeeds then each reader is totally isolated by a scientific article that links itself to masses of new resources. In practice, there is only one reversible situation, which is just the opposite of that described by Galileo: how to be 2000 against one.

(2) Positioning tactics

The more we go into this strange literature generated by controversies, the more it becomes difficult to read. This difficulty comes from the number of elements simultaneously gathered at one point – the difficulty is heightened by the acronyms, symbols and shorthand used in order to stack in the text the maximum number of resources as quickly as possible. But are numbers sufficient to convince the five or six readers left? No, of course head counts are no more sufficient in scientific texts than in war. Something more is needed: numbers must be arrayed and drilled. What I will call their **positioning** is necessary. Strangely, this is easier to understand than what we have just described since it is much closer to what is commonly called rhetoric.

(a) STACKING

Bringing pictures, figures, numbers and names into the text and then folding them is a source of strength, but it may also turn out to be a major weakness. Like references (see above Part B, section 2), they show the reader what a statement is tied to, which also means the reader knows where to pull if he or she wishes to unravel the statement. Each layer should then be carefully stacked on the former to avoid gaps. What makes this operation especially difficult is that there are indeed many gaps. The figure in excerpt (29) does not show GRF; it shows two superimposed pictures from one protocol in one laboratory in 1982; these pictures are said to be related to two tumours from one French patient in a Lyon hospital. So what is shown? GRF or meaningless scribbles on the printout of an instrument hooked up to a patient? Neither the first, nor the second. It depends on what happens to the text later on. What is shown is a stack of layers, each one *adding something* to the former. In Figure 1.7 I picture this stacking using another example. The lowest layer is made of three hamster kidneys, the highest, that is the title, claims to show 'the mammal countercurrent structure in kidney'. In dark lines I have symbolised the gain from one layer to the next. A text is like a bank; it lends more money than it has in its vault! The metaphor is a good one

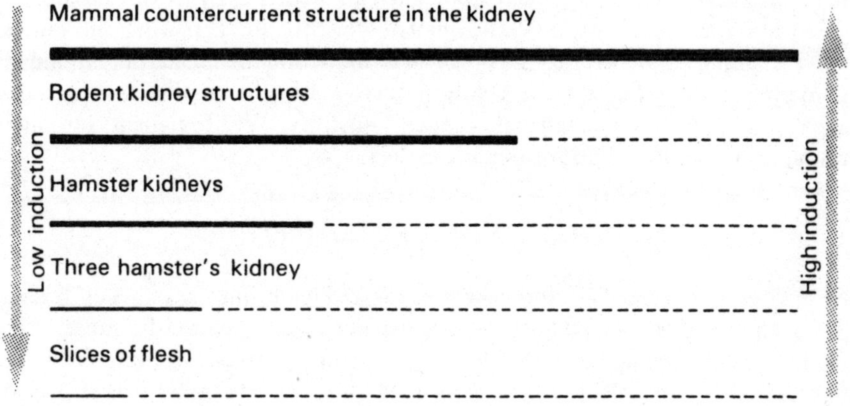

Mammal countercurrent structure in the kidney

Rodent kidney structures

Hamster kidneys

Three hamster's kidney

Slices of flesh

Low induction

High induction

Figure 1.7

since texts, like banks, may go bankrupt if all their depositors simultaneously withdraw their confidence.

If all goes well, then the article sketched in Figure 1.7 has shown mammal kidney structure; if all goes badly, it shrinks to three hamsters in one laboratory in 1984. If only a few readers withdraw their confidence, the text lingers in any of the intermediate stages: it might show hamster kidney structure, or rodent kidney structure, or lower mammal kidney structure. We recognise here the two directions in fact-building or fact-breaking that we discussed earlier.

This extreme variation between the lower and the upper layers of a paper is what philosophers often call **induction**. Are you allowed to go from a few snippets of evidence to the largest and wildest claims? From three hamsters to the mammals? From one tumour to GRF? These questions have no answer in principle since it all depends on the intensity of the controversies with other writers. If you read Schally's article now, you do not see GHRH, but a few meaningless bars and spots; his claim 'this is the GHRH structure' which was the content of sentence (5), is now seen as an empty bluff, like a cheque that bounced. On the contrary, reading Guillemin's article, you *see* GRF in the text because you believe his claim expressed in sentence (24). In both cases the belief and the disbelief are making the claim more real or less real later on. Depending on the field, on the intensity of the competition, on the difficulty of the topic, on the author's scruples, the stacking is going to be different. No matter how different the cases we could look at, the name of the game is simple enough. First rule: never stack two layers exactly one on top of the other; if you do so there is no gain, no increment, and the text keeps repeating itself. Second rule: never go straight from the first to the last layer (unless there is no one else in the field to call your bluff). Third rule (and the most important): prove as much as you can with as little as you can considering the circumstances. If you are too timid, your paper will be lost, as it will if you are too audacious. The stacking of a paper is similar to the building of a stone hut; each stone must go further than the one before. If it goes too far, the whole vault falls down; if not far enough, there will be no vault at all! The practical answers to the problem of induction are much more mundane than philosophers would wish. On these answers rests much of the strength that a paper is able to oppose to its readers' hostility. Without them, the many resources we analysed above remain useless.

(b) STAGING AND FRAMING

No matter how numerous and how well stacked its resources, an article has not got a chance if it is read just by *any* passing reader. Naturally, most of the readership has already been defined by the medium, the title, the references, the figures and the technical details. Still, even with the remainder it is still at the mercy of malevolent readers. In order to defend itself the text has to explain how and by whom it should be read. It comes, so to speak, with its own user's notice, or legend.

The image of the ideal reader built into the text is easy to retrieve. Depending on the author's use of language, you immediately imagine to whom he or she is talking (at least you realise that in most cases he or she is not talking to you!). Sentence (24), that defined the amino acid structure of GRF, is not aimed at the same reader as the following:

> (31) There exists a substance that regulates body growth; this substance is itself regulated by another one, called GRF; it is made of a string of 44 amino acids (amino acids are the building block of all proteins); this string has recently been discovered by the Nobel Prize winner Roger Guillemin.

Such a sentence is addressed to a completely different audience. More people are able to read it than sentence (24) or (26). *More* people but equipped with *fewer* resources. Notice that popularisation follows the same route as controversy but in the opposite direction; it was because of the intensity of the debates that we were slowly led from non-technical sentences, from large numbers of ill-equipped verbal contestants to small numbers of well-equipped contestants who write articles. If one wishes to increase the number of readers again, one has to decrease the intensity of the controversy, and reduce the resources. This remark is useful because the difficulty of writing 'popular' articles about science is a good measure of the accumulation of resources in the hands of few scientists. It is hard to popularise science because it is designed to force out most people in the first place. No wonder teachers, journalists and popularisers encounter difficulty when we wish to bring the excluded readership back in.

The kind of words authors use is not the only way of determining the ideal reader at whom they are aiming. Another method is to anticipate readers' objections in advance. This is a trick common to all rhetoric, scientific or not. 'I knew you would object to this, but I have already thought of it and this is my answer.' The reader is not only chosen in advance, but what it is going to say is taken out of its own mouth, as I showed for instance in excerpt (27) (I use 'it' instead of 'he or she' because this reader is not a person in the flesh but a person on paper, a **semiotic character**).[12] Thanks to this procedure, the text is carefully aimed; it exhausts all potential objections in advance and may very well leave the reader speechless since it can do nothing else but take the statement up as a matter of fact.

What sort of objections should be taken into account by the author? Again, this is a question that philosophers try to answer in principle although it only has practical answers, depending on the battlefield. The only rule is to ask the (imaginary) reader what sort of **trials** it will require before believing the author. The text builds a little story in which something incredible (the hero) becomes gradually more credible because it withstands more and more terrible trials. The implicit dialogue between authors and readers then takes something of this form:

> (34)–If my substance triggers growth hormone in three different assays, will you believe it to be GRF?
> –No, this is not enough, I also want you to show me that your stuff from a pancreas tumour is the same as the genuine GRF from the hypothalamus.
> –What do you mean 'the same'; what trials should my stuff, as you say, undergo to be called 'genuine GRF'?
> – The curves of your stuff from the pancreas and GRF from the hypothalamus should be superimposed; this is the trial I want to see with my own eyes before I believe you. I won't go along with you without it.
> –This is what you want? And after that you give up? You swear? Here it is: see figure 2, perfect superimposition!
> –Hold on! Not so fast! This is not fair; what did you do with the curves to get them to fit?
> –Everything that could be done given the present knowledge of statistics and today's computers. The lines are theoretical, computer-calculated and drawn, from the four-parameter logistic equations for each set of data! Do you give up now?
> –Yes, yes, certainly, I believe you!

'It' gives up, the imaginary reader whose objections and requirements have been anticipated by the master author!

Scientific texts look boring and drab from the most superficial point of view. If the reader recomposes the challenge they take up, they are as thrilling as story telling. 'What is going to happen to the hero? Is it going to resist this new ordeal? No, it is too much even for the best. Yes, it did win? How incredible. Is the reader convinced? Not yet. Ah hah, here is a new test; impossible to meet these requirements, too tough. Unfair, this is unfair.' Imagine the cheering crowds and the boos. No character on stage is watched with such passion and asked to train and rehearse as is, for instance, this GRF stuff.

The more we get into the niceties of the scientific literature, the more extraordinary it becomes. It is now a real opera. Crowds of people are mobilised by the references; from offstage hundreds of accessories are brought in. Imaginary readers are conjured up which are not asked only to believe the author but to spell out what sort of tortures, ordeals and trials the heroes should undergo before being recognised as such. Then the text unfolds the dramatic story of these trials. Indeed, the heroes triumph over all the powers of darkness, like the Prince in *The Magic Flute*. The author adds more and more impossible trials just, it seems, for the pleasure of watching the hero overcoming them. The authors challenge the audience and their heroes sending a new bad guy, a storm, a devil, a curse, a dragon, and the heroes fight them. At the end, the readers, ashamed of their former doubts, have to accept the author's claim. These operas unfold thousands of times in the pages of *Nature* or the *Physical Review* (for the benefit, I admit, of very, very few spectators indeed).

The authors of scientific texts do not merely build readers, heroes and trials into the paper. They also make clear who they are. The authors in the flesh become the authors on paper, adding to the article more semiotic characters, more 'its'. The six authors of what I called Guillemin's paper did not, of course, write it. No one could remember how many drafts the paper passed through. The attribution of these six names, the order in which they enter, all that is carefully staged, and since this is one part of the writing of the plot, it does not tell us *who* writes the plot.

This obvious staging is not the only sign of the authors' presence. Although technical literature is said to be impersonal, this is far from being so. The authors are everywhere, built into the text. This can be shown even when the passive voice is used – this trait being often invoked to define scientific style. When you write: 'a portion of tissue from each tumour was extracted, a picture of the author is drawn as much as if you write 'Dr Schally extracted' ' or 'my young colleague Jimmy extracted'. It is just another picture; a grey backdrop on a stage is as much a backdrop as a coloured one. It all depends on the effects one wishes to have on the audience.

The portrayal of the author is important because it provides the imaginary counterpart of the reader; it is able to control how the reader should read, react and believe. For instance, it often positions itself in a genealogy which already presages the discussion:

> (33) Our conception of the hamster kidney structure has recently been dramatically altered by Wirz's observations (reference). We wish to report a new additional observation.

The author of this sentence does not portray itself as a revolutionary, but as a follower; not as a theoretician, but as a humble observer. If a reader wishes to attack the claim or the theory, it is redirected to the 'dramatic' transformations Wirz made and to the 'conceptions' he had. To show how such a sentence makes up a certain image of the author, let us rewrite it:

> (34) Wirz (reference) recently observed a puzzling phenomenon he could not interpret within the classical framework of kidney structure. We wish to propose a new interpretation of his data.

The article has immediately changed tack. It is now a revolutionary article and a theoretical one. Wirz's position has been altered. He was the master; he is now a precursor who did not know for sure what he was doing. The reader's expectations will be modified depending on which version the author chooses. The same changes will occur if we fiddle with sentence (21), which was the introduction to the paper written by Guillemin to announce the discovery of GRF. Remember that Schally's earlier endeavours were dismissed with the sentence: 'so far, hypothalamic GRF has not been unequivocally characterised, despite earlier claims to the contrary'. What does the reader feel if we now transform sentence (21) into this one:

> (35) Schally (reference) earlier proposed a characterisation of hypothalamic GRF; the present work proposes a different sequence which might solve some of the difficulties of the former characterisation.

The reader of sentence (21) is expecting truth at last after many senseless attempts at finding GRF, whereas the reader of (35) is prepared to read a new tentative proposition that situates itself in the same lineage as the former. Schally is a nonentity in the first case, an honourable colleague in the second. Any change in the author's position in the text may modify the readers' potential reactions.

Especially important is the staging by the author of what should be discussed, what is really interesting (what is especially important!) and what is, admittedly, disputable. This hidden agenda, built into the text, paves the way for the discussion. For instance, Schally, at the end of the article that I have used all along as an example, is suddenly not sure of anything any more. He writes:

> (36) Whether this molecule represents the hormone which is responsible for the stimulation of growth hormone released under physiological conditions can only be proven by further studies.

This is like taking out an insurance policy against the unexpected transformation of facts into artefacts. Schally did not say that he found 'the' GHRH, but only 'a' molecule that looked like GHRH. Later on, when he was so violently criticised for his blunder, he was then able to say that he never claimed that GHRH was the molecule cited in claim (5).

This caution is often seen as the sign of scientific style. Understatement would then be the rule and the difference between technical literature and literature in general would be the multiplication of negative modalities in the former. We now know this to be as absurd as saying that one walks only with one's left leg. Positive modalities are as necessary as negative ones. Each author allocates what shall not be discussed and what ought to be discussed (see again (21)). When it is necessary not to dispute a black box there is no understatement whatsoever. When the author is on dangerous ground, understatement proliferates. Like all the effects we have seen in this section, it all depends on circumstances. It is impossible to say that technical literature always errs on the side of caution; it also errs on the side of audacity; or rather it does not err, it zigzags through obstacles, and evaluates the risks as best it can. Guillemin, for instance, at the end of his paper runs hot and cold at the same time:

(39) What can certainly be said is that the molecule we have now characterized has all the attributes expected from the long-sought hypothalamic releasing factor for growth hormone.

Schally's caution is gone. The risk is taken; certainty is on their side: the new substance does everything that GRF does. The author simply stops short of saying 'this *is* GRF'. (Note that the author happily uses 'we' and the active voice when summarising its victory.) But the next paragraph adopts entirely different tactics:

(38) In keeping with other past experience, probably the most interesting role, effect, or use of GRF is currently totally unsuspected.

This is indeed an insurance policy against the unknown. No one will be able to criticise the author for its lack of vision, since the unexpected is expected. By using such a formula, the author protects itself against what happened in the past with another substance, somatostatin.[13] Originally isolated in the hypothalamus to inhibit the release of growth hormone, it turned out to be in the pancreas and to play a role in diabetes. But Guillemin's group missed this discovery that others made with their own substance. So, is the author cautious or not? Neither. It carefully writes to protect its claims as best as it can and to fence off the reader's objections.

Once a paper is written, it is very difficult to retrieve the careful tactics through which it was crafted, although a look at the drafts of scientific articles will be enough to show that the real authors are quite self-conscious about all of this. They know that without rewriting and positioning, the strength of their paper will be spoiled, because the authors and the readers built into the text do not match. Everything is at the mercy of a few ill-chosen words. The claim may become wild, the paper controversial, or, on the contrary, so timid and over-cautious, so polite and tame that it lets others reap the major discoveries.

(c) CAPTATION

It may be discouraging for those of us who want to write powerful texts able to influence controversies, but even the enormous amount of work shown above is not enough! Something is still missing. No matter how many references the author has been able to muster; no matter how many resources, instruments and pictures it has been capable of mobilising in one place; no matter how well arrayed and drilled its troops are; no matter how clever its anticipation of what the readers will do and how subtle the presentation of itself; no matter how ingenious the choice of which ground should be held and which may be abandoned; regardless of all these strategies, the real reader, the reader in the flesh, the 'he' or 'she' may still *reach different conclusions*. Readers are devious people, obstinate and unpredictable – even the five or six left to read the paper from beginning to end. Isolated, surrounded, besieged by all your allies, they can still escape and conclude that Soviet missiles are accurate to within 100 metres, that you have not proven the existence of GHRH or GRF, or that your paper on fuel cells is a mess. The paper-reader, the 'it' of, for instance, statement (32), may have stopped discussing and admitted the writer's credibility; but what about the real reader? He or she might have skipped a passage entirely, focused on a detail marginal to the author. The author told them in claim (21) that hypothalamus control of growth hormone is indisputable: are they going to follow him? It told them in (36) what was to be discussed; are they going to accept this agenda? The writer draws so many pathways going from one place to another and asks the reader to follow them; the readers may cross these paths and then escape. To come back to Galileo's sentence, 2000 Demosthenes and Aristotles are still weak if one average reader is allowed to break away and flee. All the numbers amassed by the technical literature are not enough if the reader is allowed to stroll and wander. All the objectors' moves should then be controlled so that they encounter massive numbers and are defeated. I call **captation** (or captatio in the old rhetoric) this subtle control of the objectors' moves.[14]

Remember that the authors need the readers' willingness to have their own claims turned into facts (see Part A, section 2). If the readers are put off, they are not going to take up the claim; but if they are left free to discuss the claim, it will be deeply altered. The writer of a scientific text is then in a quandary: how to leave someone completely free and have them at the same time completely obedient. What is the best way to solve this paradox? To lay out the text so that wherever the reader is there is only *one way* to go.

But how can this result be achieved, since by definition the real reader may dispute everything and go in any direction? By making it more difficult for the reader to go in all the other directions. How can this be achieved? By carefully stacking more black boxes, less easily disputable arguments. The nature of the game is exactly like that of building a dam. It would be foolish for a dam engineer to suppose that the water will obey his wishes, abstaining from overflowing or politely running from bottom to top. On the contrary, any engineer should start with the principle that if water can leak away it will. Similarly with readers, if you leave the smallest outlet open to them they will rush out; if you try to force them to go upstream they will not. So what you have to do is to make sure the reader always flows freely but *in a deep enough valley*! Since the beginning of this chapter we have observed this digging, trenching and damming many times over. All the examples moved from a better-known statement to a lesser-known one; all were using a less easily disputable claim to start or to stop discussion on a statement easier to dispute. Each controversy aimed at reversing the flow by shifting negative and positive modalities. Captation is a generalisation of the same phenomenon inducing readers to move far away from what they were ready to accept at first. If the digging and damming is well set up, the reader, although taken in, will feel entirely free (see Figure 1.8).

The hydraulic metaphor is an apt one since the scale of public work to be undertaken depends on how far you wish to force the water to go, on the intensity of the flow, on the slope and on what kind of landscape you have to buttress the dams and the ducts. It is the same thing with persuasion. It is an easy job if you

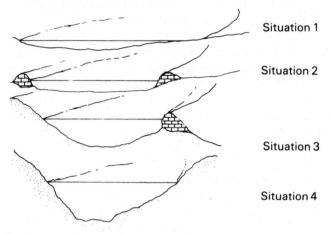

Situation 1

Situation 2

Situation 3

Situation 4

From unconvincing to incontrovertible evidence...

Figure 1.8

want to convince a few people of something that is almost obvious; it is much harder if you wish to convince a large number of people of something very remote from or even contrary to their current beliefs (see Chapter 5, Part C). This metaphor shows that the relation between the amount of work and persuasion depends on the circumstances. Convincing is not just a matter of throwing words about. It is a race between the authors and the readers to control each other's moves. It would be enormously difficult for one 'average man' to force off their paths '2000 Demosthenes and Aristotles' in a matter where, at first sight, every direction is equally possible; the only way to decrease the difficulty is to dam up all the alternative channels. No matter where the reader is in the text, he or she is confronted with instruments harder to discuss, figures more difficult to doubt, references that are harder to dispute, arrays of stacked black boxes. He or she flows from the introduction to the conclusion like a river flowing between artificial banks.

When such a result is attained – it is very rare – a text is said to be **logical**. Like the words 'scientific' or 'technical', it seems that 'logical' often means a different literature from the illogical type that would be written by people with different kinds of minds following different methods or more stringent standards. But there is no absolute break between logical and illogical texts; there is a whole gamut of nuances that depend as much on the reader as on the author. Logic refers not to a new subject matter but to simple practical schemes: Can the reader get out? Can he easily skip this part? Is she able, once there, to take another path? Is the conclusion escapable? Is the figure waterproof? Is the proof tight enough? The writer arrays whatever is at hand in tiers so that these questions find practical answers. This is where **style** starts to count; a good scientific writer may succeed in being 'more logical' than a bad one.

The most striking aspect of this race between the reader and the writer is when the limits are reached. In principle, of course, there is no limit since the fate of the statement is, as I said, in later users' hands (see Chapter 2, Part C). It is always possible to discuss an article, an instrument, a figure; it is always possible for a reader-in-the-flesh to move off the path expected of the reader-in-the-text. In practice, however, limits are reached. The author obtains this result by stacking so many tiers of black boxes that at one point the reader, obstinate enough to dissent, will be confronted with facts so old and so unanimously accepted that in order to go on doubting he or she will be *left alone*. Like a clever engineer who decides to build her dam on solid bedrock, the writer will manage to link the fate of the article to that of harder and harder facts. The practical limit is reached when the average dissenter is no longer faced with the author's opinion but with what thousands and thousands of people have thought and asserted. Controversies have an end after all. The end is not a natural one, but a carefully crafted one like those of plays or movies. If you still doubt that the MX should be built (see (1)), or that GHRH has been discovered by Schally (see (5)), or that fuel cells are the future of the electric engine (see (8)), then you will be all by yourself, without support and ally, alone in your profession, or, even worse, isolated from the community, or maybe, still more awful, sent to an asylum! It is a powerful rhetoric that which is able to drive the dissenter mad.

(3) The second rule of method

In this chapter we have learned a **second rule of method** in addition to the first one that required us to study science and technology in action. This second rule asks us not to look for the intrinsic qualities of any given statement but to look instead for all the transformations it undergoes later in other hands. This rule is the consequence of what I called our first principle: the fate of facts and machines is in the hands of later users.

These two rules of method taken together allow us to continue our trip through technoscience without being intimidated by the technical literature. No matter what controversy we start from, we will always be able to take our bearings:

(a) by looking at the stage the claim we chose as our departure point is at;
(b) by finding the people who are striving to make this claim more of a fact and those who are trying to make it less of a fact;
(c) by checking in which direction the claim is pushed by the opposite actions of these two groups of people; is it up the ladder drawn in Figure 1.5 or down?

This initial enquiry will give us our first bearing (our latitude so to speak). Then, if the statement we follow is quickly destroyed, we will have to see how it is transformed and what happens to its new version: is it more easily accepted or less? The new enquiry will offer us:

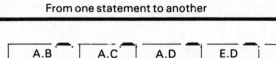

Figure 1.9

(d)   a measure of the distance between the original claim and the new ones, as we saw for instance between Schally's sentence (5) about GHRH made in 1971, and Guillemin's claim made in 1982 about the same substance named GRF and with a completely different amino acid sequence. This drift will provide us with our second bearing, our longitude.

Finally, the two dimensions put together will draw:

(e)   the front line of the controversy as summarized in Figure 1.9.

Conclusion
Numbers, more numbers

Having reached the end of this chapter, it should be clear now why most people do not write and do not read scientific texts. No wonder! It is a peculiar trade in a merciless world. Better read novels! What I will call **fact-writing** in opposition to fiction-writing limits the number of possible readings to three: giving up, going along, working through. *Giving up* is the most usual one. People give up and do *not* read the text, whether they believe the author or not, either because they are pushed out of the controversy altogether or because they are not interested in reading the article (let us estimate this to be 90 per cent of the time). *Going along* is the rare reaction, but it is the normal outcome of scientific rhetoric: the reader believes the author's claim and helps him to turn it into a fact by using it further with no dispute (maybe 9 per cent of the time?). There is still one more possible outcome, but such a rare and costly one that it is almost negligible as far as numbers are concerned: *re-enacting* everything that the authors went through. This last issue remains open because there is always at least one flaw even in the best written scientific text: many resources mobilised in it are said to come from instruments, animals, pictures, from things *out of the text*. The adamant objector could then try to put the text in jeopardy by untying these supply lines. He or she will then be led from the text to where the text claims to come from: Nature or the laboratory. This is possible on one condition: that the dissenter is equipped with a laboratory or with ways to get straight at Nature more or less similar to that of the author. No wonder this way of reading a scientific paper is rare! You have to have a whole machinery of your own. Resuming the controversy, reopening the black box is achieved at this price, and only at this price. It is this rare remaining strategy that we will study in the next chapter.

The peculiarity of the scientific literature is now clear: the only three possible readings all lead to the demise of the text. If you give up, the text does not count and might as well not have been written at all. If you go along, you believe it so much that it is quickly abstracted, abridged, stylised and sinks into tacit practice. Lastly, if you work through the authors' trials, you quit the text and enter the laboratory. Thus the scientific text is chasing its readers away whether or not it is successful. Made for attack and defence, it is no more a place for a leisurely stay than a bastion or a bunker. This makes it quite different from the reading of the Bible, Stendhal or the poems of T.S. Eliot.

Yes, Galileo was quite mistaken when he purported to oppose rhetoric and science by putting big numbers on one side and one 'average man' who happened to 'hit upon the truth' on the other. Everything we have seen since the beginning indicates exactly the opposite. Any average man starting off a dispute ends up being confronted with masses of resources, not just 2000, but tens of thousands. So what is the difference between rhetoric, so much despised, and science, so much admired? Rhetoric used to be despised because it mobilised *external allies* in favour of an argument, such as passion, style, emotions, interests, lawyers' tricks and so on. It has been hated since Aristotle's time because the regular path of reason was unfairly distorted or reversed by any passing sophist who invoked passion and style. What should be said of the people who invoke so many more external allies besides passion and style in order to reverse the path of common reasoning? The difference between the old rhetoric and the new is not that the first makes use of external allies which the second refrains from using; the difference is that the first uses only *a few* of them and the second *very many*. This distinction allows me to avoid a wrong way of interpreting this chapter which would be to say that we studied the 'rhetorical aspects' of technical literature, as if the other aspects could be left to reason, logic and technical details. My contention is that on the contrary we must eventually come to call scientific the rhetoric able to mobilise on one spot more resources than older ones (see Chapter 6).

It is because of this definition in terms of the number of allies that I abstained from defining this literature by its most obvious trait: the presence of numbers, geometrical figures, equations, mathematics, etc. The presence of these objects will be explained only in Chapter 6 because their form is impossible to understand when separated from this mobilisation process made necessary by the intensity of the rhetoric. So the reader should not be worried either by the presence or by the absence of figures in the technical literature. So far it is not the relevant feature. We have to understand first how many elements can be brought to bear on a controversy; once this is understood, the other problems will be easier to solve.

By studying in this chapter how a controversy gets fiercer, I examined the anatomy of technical literature and I claimed that it was a convenient way to make good my original promise to show the heterogeneous components that make up technoscience, including the *social* ones. But I'd rather anticipate the objection of my (semiotic) reader: 'What do you mean "social"?' it indignantly says. 'Where is capitalism, the proletarian classes, the battle of the sexes, the struggle for the emancipation of the races, Western culture, the strategies of wicked multinational corporations, the military establishment, the devious interests of professional lobbies, the race for prestige and rewards among scientists? All these elements are social and this is what you did *not show* with all your texts, rhetorical tricks and technicalities!'

I agree, we saw nothing of that sort. What I showed, however, was something much more obvious, much less far-fetched, much more pervasive than any of these traditional social actors. We saw a literature becoming more technical by bringing in more and more resources. In particular, we saw a dissident driven into isolation because of the number of elements the authors of scientific articles mustered on their side. Although it sounds counter-intuitive at first, the more technical and specialised a literature is, the more 'social' it becomes, since the *number of associations* necessary to drive readers out and force them into accepting a claim as a fact increase. Mr Anybody's claim was easy to deny; it was much harder to shrug off Schally's article on GHRH, sentence (16), not because the first is social and the second technical, but because the first is one man's word and the second is many well-equipped men's words; the first is made of a few associations, the second of many. To say it more bluntly, the first is a little social, the second *extremely* so. Although this will become understandable much later, it is already clear that if being isolated, besieged, and left without allies and supporters is not a social act, then nothing is. The distinction between the technical literature and the rest is not a natural boundary; it is a border created by the disproportionate amount of linkages, resources and allies locally available. This literature is so hard to read and analyse not because it escapes from all normal social links, but because it is *more* social than so-called normal social ties.

CHAPTER 2

Laboratories

We could stop our enquiry where we left it at the end of the previous chapter. For a layperson, studying science and technology would then mean analysing the discourse of scientists, or counting citations, or doing various bibliometric calculations, or performing semiotic studies[1] of scientific texts and of their iconography, that is, extending literary criticism to technical literature. No matter how interesting and necessary these studies are, they are not sufficient if we want to follow scientists and engineers at work; after all, they do not draft, read and write papers twenty-four hours a day. Scientists and engineers invariably argue that there is something behind the technical texts which is much more important than anything they write.

At the end of the previous chapter, we saw how the articles forced the reader to choose between three possible issues: giving up (the most likely outcome), going along, or working again through what the author did. Using the tools we devised in Chapter 1, it is now easy to understand the first two issues, but we are as yet unable to understand the third. Later, in the second part of this book, we will see many other ways to avoid this issue and still win over in the course of a controversy. For the sake of clarity, however, I make the supposition in this part that the dissenter has no other escape but to work through what the author of the paper did. Although it is a rare outcome, it is essential for us to visit the places where the papers are said to originate. This new step in our trip through technoscience is much more difficult, because, whilst the technical literature is accessible in libraries, archives, patent offices or corporate documentation centres, it is much less easy to sneak into the few places where the papers are written and to follow the construction of facts in their most intimate details. We have no choice, however, if we want to apply our first rule of method: if the scientists we shadow go inside laboratories, then we too have to go there, no matter how difficult the journey.

Part A
From texts to things: a showdown

'You doubt what I wrote? Let me show you.' The very rare and obstinate dissenter who has *not* been convinced by the scientific text, and who has not found other ways to get rid of the author, is led from the text into the place where the text is said to come from. I will call this place the **laboratory**, which for now simply means, as the name indicates, the place where scientists *work*. Indeed, the laboratory was present in the texts we studied in the previous chapter: the articles were alluding to 'patients', to 'tumours', to 'HPLC', to 'Russian spies', to 'engines'; dates and times of experiments were provided and the names of technicians acknowledged. All these allusions however were made within a paper world; they were a set of semiotic actors presented in the text but not *present* in the flesh; they were alluded to as if they existed independently from the text; they could have been invented.

(1) Inscriptions

What do we find when we pass through the looking glass and accompany our obstinate dissenter from the text to the laboratory? Suppose that we read the following sentence in a scientific journal and, for whatever reason, do not wish to believe it:

> (1) 'Fig.1 shows a typical pattern. Biological activity of endorphin was found essentially in two zones with the activity of zone 2 being totally reversible, or statistically so, by naloxone.'

We, the dissenters, question this figure 1 so much, and are so interested in it, that we go to the author's laboratory (I will call him 'the Professor'). We are led into an air-conditioned, brightly lit room. The Professor is sitting in front of an array of devices that does not attract our attention at first. 'You doubt what I wrote? Let me show you.' This last sentence refers to an image slowly produced by one of these devices (Figure 2.1):

(2)

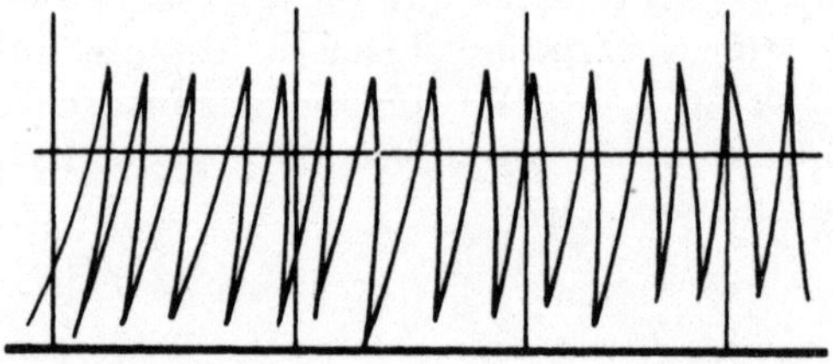

Figure 2.1

'OK. This is the base line; now, I am going to inject endorphin, what is going to happen? See?!' (Figure 2.2)

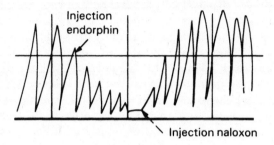

Figure 2.2

'Immediately the line drops dramatically. And now watch naloxone. See?! Back to base line levels. It is fully reversible.'

We now understand that what the Professor is asking us to watch is related to the figure in the text of sentence (1). We thus realise where this figure comes from. It has been *extracted* from the instruments in this room, *cleaned, redrawn,* and *displayed.* We now seem to have reached the source of all these images that we saw arrayed in the text as the final proofs of all the arguments in Chapter 1. We also realise, however, that the images that were the last layer in the text, are the *end result* of a long process in the laboratory that we are now starting to observe. Watching the graph paper slowly emerging out of the physiograph, we understand that we are at the junction of two worlds: a paper world that we have just left, and one of instruments that we are just entering. A hybrid is produced at the interface: a raw image, to be used later in an article, that is emerging from an instrument.

For a time we focus on the stylus pulsating regularly, inking the paper, scribbling cryptic notes. We remain fascinated by this fragile film that is in between text and laboratory. Soon, the Professor draws our attention beneath and beyond the traces on the paper, to the physiograph from which the image is slowly being emitted. Beyond the stylus a massive piece of electronic hardware records, calibrates, amplifies and regulates signals coming from another instrument, an array of glassware. The Professor points to a glass chamber in which bubbles are regularly flowing around a tiny piece of something that looks like elastic. It is indeed elastic, the Professor intones. It is a piece of gut, guinea pig gut ('myenteric plexus-longitudinal muscle of the guinea pig ileum', are his words). This gut has the property of contracting regularly if maintained alive. This regular pulsation is easily disturbed by many chemicals. If one hooks the gut up so that each contraction sends out an electric pulse, and if the pulse is made to move a stylus over graph paper, then the guinea pig gut will be induced to produce regular scribbles over a long period. If you then add a chemical to the chamber you *see* the peaks drawn by the inked stylus slow down or accelerate at the other end. This perturbation, invisible in the chamber, is visible on paper: the chemical, no matter what it is, is given a *shape* on paper. This shape 'tells you something' about the chemical. With this set-up you may now ask new questions: if I double the dose of chemical will the peaks be doubly decreased? And if I triple it, what will happen? I can now measure the white surface left by the decreasing scribbles directly on the graph paper, thereby defining a quantitative relation between the dose and the response. What if, just after the first chemical is added, I add another one which is known to counteract it? Will the peaks go back to normal? How fast will they do so? What will be the pattern of this return to the base line level? If two chemicals, one known, the other unknown, trace the same slope on the paper, may I say, in this respect at least, that they are the same chemicals? These are some of the questions the Professor is tackling with endorphin (unknown), morphine (well known) and naloxone (known to be an antagonist of morphine).

We are no longer asked to believe the text that we read in *Nature*; we are now asked to believe *our own eyes*, which can see that endorphin is behaving exactly like morphine. The object we looked at in the text and the one we are now contemplating are identical except for one thing. The graph of sentence (1), which was the most concrete and visual element of the text, is now in (2) the most abstract and textual element in a bewildering array of equipment. Do we see more or less than before? On the one hand we can see more, since we are looking at not only the graph but also the physiograph, and the electronic hardware, and the glassware, and the electrodes, and the bubbles of oxygen, and the pulsating ileum, and the Professor who is injecting chemicals into the chamber with his syringe, and is writing down in a huge protocol book the time, amount of and reactions to the doses. We can see more, since we have before our eyes not only the image but what the image is made of.

On the other hand we see *less* because now each of the elements that makes up the final graph could be modified so as to produce a different visual outcome. Any number of incidents could blur the tiny peaks and turn the regular writing into a meaningless doodle. Just at the time when we feel comforted in our belief and start to be fully convinced by our own eyes watching the image, we suddenly feel uneasy because of the fragility of the whole set up. The Professor, for instance, is swearing at the gut saying it is a 'bad gut'. The technician who sacrificed the guinea pig is held responsible and the Professor decides to make a fresh start with a new animal. The demonstration is stopped and a new scene is set up. A guinea pig is placed on a table, under surgical floodlights, then anaesthetised, crucified and sliced open. The gut is located, a tiny section is extracted, useless tissue peeled away, and the precious fragment is delicately hooked up between two electrodes and immersed in a nutrient fluid so as to be maintained alive. Suddenly, we are much further from the paper world of the article. We are now in a puddle of blood and viscera, slightly nauseated by the extraction of the ileum from this little furry creature. In the last chapter, we admired the rhetorical abilities of the Professor as an author. Now, we realise that many other manual abilities are required in order to write a convincing paper later on. The guinea pig alone would not have been able to tell us anything about the similarity of endorphin to morphine; it was not mobilisable into a text and would not help to convince us. Only a part of its gut, tied up in the glass chamber and hooked up to a physiograph, can be mobilised in the text and add to our conviction. Thus, the Professor's art of convincing his readers must extend beyond the paper to preparing the ileum, to calibrating the peaks, to tuning the physiograph.

After hours of waiting for the experiment to resume, for new guinea pigs to become available, for new endorphin samples to be purified, we realise that the invitation of the author ('let me show you') is not as simple as we thought. It is a slow, protracted and complicated staging of tiny images in front of an audience. 'Showing' and 'seeing' are not simple flashes of intuition. Once in the lab we are not presented outright with the real endorphin whose existence we doubted. We are presented with another world in which it is necessary to prepare, focus, fix and rehearse the vision of the real endorphin. We came to the laboratory in order to settle our doubts about the paper, but we have been led into a labyrinth.

This unexpected unfolding makes us shiver because it now dawns on us that if we disbelieve the traces obtained on the physiograph by the Professor, we will have to give up the topic altogether or go through the same experimental chores all over again. The stakes have increased enormously since we first started reading scientific articles. It is not a question of reading and writing back to the author any more. In order to argue, we would now need the manual skills required to handle the scalpels, peel away the guinea pig ileum, interpret the decreasing peaks, and so on. Keeping the controversy alive has already forced us through many difficult moments. We now realise that what we went through is nothing compared to the scale of what we have to undergo if we wish to continue. In Chapter 1, we only needed a good library in order to dispute texts. It might have been costly and not that easy, but it was still feasible. At this present point, in order to go on, we need guinea pigs, surgical lamps and tables, physiographs, electronic hardware, technicians and morphine, not to mention the scarce flasks of purified endorphin; we also need the skills to use all these elements and to turn them into a pertinent objection to the Professor's claim. As will be made clear in Chapter 4, longer and longer detours will be necessary to find a laboratory, buy the equipment, hire the technicians and become acquainted with the ileum assay. All this work just to start making a convincing counter-argument to the Professor's original paper on endorphin. (And when we have made this detour and finally come up with a credible objection, where will the Professor be?)

When we doubt a scientific text we do not go from the world of literature to Nature as it is. Nature is not directly beneath the scientific article; it is there *indirectly* at best (see Part C). Going from the paper to the laboratory is going from an array of rhetorical resources to a set of new resources devised in such a way as to provide the literature with its most powerful tool: the visual display. Moving from papers to labs is moving from literature to convoluted ways of getting this literature (or the most significant part of it).

This move through the looking glass of the paper allows me to define an **instrument**, a definition which will give us our bearings when entering any laboratory. I will call an instrument (or **inscription device**) any set-up, no matter what its size, nature and cost, that provides a visual display of any sort in a scientific text. This definition is simple enough to let us follow scientists' moves. For instance an optical telescope is an instrument, but so is an array of several radio-telescopes even if its constituents are separated by thousands of kilometers. The guinea pig ileum assay is an instrument even if it is small and cheap compared to an array of radiotelescopes or the Stanford linear accelerator. The definition is not provided by the cost nor by the sophistication but only by this characteristic: the set-up provides an inscription that is used as the final layer in a scientific text. An instrument, in this definition, is not every set-up which ends with a little window that allows someone to take a reading. A thermometer, a watch, a Geiger counter, all provide readings but are not considered as instruments as long as these readings are not used as the final layer of technical papers (but see Chapter 6). This point is important when watching complicated contrivances with hundreds of intermediary readings taken by dozens of white-coated technicians. What will be used as visual proof in the article will be the few lines in the bubble chamber and not the piles of printout making the intermediate readings.

It is important to note that the use of this definition of instrument is a relative one. It depends on time. Thermometers *were* instruments and very important ones in the eighteenth century, so were Geiger counters between the First and Second World Wars. These devices provided crucial resources in papers of the time. But now they are only parts of larger set-ups and are only used so that a new visual proof can be displayed at the end. Since the definition is relative to the use made of the 'window' in a technical paper, it is also relative to the intensity and nature of the associated controversy. For instance, in the guinea pig ileum assay there is a box of electronic hardware with many readings that I will call 'intermediate' because they do not constitute the visual display eventually put to use in the article. It is unlikely that anyone will quibble about this because the calibration of electronic signals is now made through a black box produced industrially and sold by the thousand. It is a different matter with the huge tank built in an old gold mine in South Dakota at a cost of $600,000 (1964 dollars!) by Raymond Davis[2] to detect solar neutrinos. In a sense the whole set-up may be considered as *one* instrument providing one final window in which astro-physicists can read the number of neutrinos emitted by the sun. In this case all the other readings are intermediate ones. If the controversy is fiercer, however, the set-up is broken down into *several* instruments, each providing a specific visual display which has to be independently evaluated. If the controversy heats up a bit we do not see neutrinos coming out of the sun. We see and hear a Geiger counter that clicks when Argon[37] decays. In this case the Geiger counter, which gave only an intermediate reading when there was no dispute, becomes an instrument in its own right when the dispute is raging.

The definition I use has another advantage. It does not make presuppositions about what the instrument is made of. It can be a piece of hardware like a telescope, but it can also be made of softer material. A statistical institution that employs hundreds of pollsters, sociologists and computer scientists gathering all sorts of data on the economy *is* an instrument if it yields inscriptions for papers written in economic journals with, for instance, a graph of the inflation rate by month and by branch of industry. No matter how many people were made to participate in the construction of the image, no matter how long it took, no matter how much it cost, the whole institution is used as *one* instrument (as long as there is no controversy that calls its intermediate readings into question).

At the other end of the scale, a young primatologist who is watching baboons in the savannah and is equipped only with binoculars, a pencil and a sheet of white paper may be seen as an instrument if her coding of baboon behaviour is summed up in a graph. If you want to deny her statements, you might (everything else being equal) have to go through the same ordeals and walk through the savannah taking notes with similar constraints. It is the same if you wish to deny the inflation rate by month and industry, or the detection of endorphin with the ileum assay. The instrument, whatever its nature, is what leads you from the paper to what supports the paper, from the many resources mobilised in the text to the many more resources mobilised to create the visual displays of the texts. With this definition of an instrument, we are able to ask many questions and to make comparisons: how expensive they are, how old they are, how many intermediate readings compose one instrument, how long it takes to get one reading, how many people are mobilised to activate them, how many authors are using the inscriptions they provide in their papers, how controversial are those readings . . . Using this notion we can define more precisely than earlier the laboratory as any place that gathers one or several instruments together.

What is behind a scientific text? Inscriptions. How are these inscriptions obtained? By setting up instruments. This other world just beneath the text is invisible as long as there is no controversy. A picture of moon valleys and mountains is presented to us as if we could see them directly. The telescope that makes them visible is invisible and so are the fierce controversies that Galileo had to wage centuries ago to produce an image of the Moon. Similarly, in Chapter 1, the accuracy of Soviet missiles was just an *obvious* statement; it became the outcome of a complex system of satellites, spies, Kremlinologists and computer simulation, only *after* the controversy got started. Once the fact is constructed, there is no instrument to take into account and this is why the painstaking work necessary to tune the instruments often disappears from popular science. On the contrary, when science in action is followed, instruments become the crucial elements, immediately after the technical texts; they are where the dissenter is inevitably led.

There is a corollary to this change of relevance on the inscription devices depending on the strength of the controversy, a corollary that will become more important in the next chapter. If you consider only fully-fledged facts it seems that everyone could accept or contest them equally. It does not cost anything to contradict or accept them. If you dispute further and reach the frontier where facts are made, instruments become visible and with them the cost of continuing the discussion rises. It appears that *arguing is costly*. The equal world of citizens having opinions about things becomes an unequal world in which dissent or consent is not possible without a huge accumulation of resources which permits the collection of relevant inscriptions. What makes the differences between author and reader is not only the ability to utilise all the rhetorical resources studied in the last chapter, but also to gather the many devices, people and animals necessary to produce a visual display usable in a text.

(2) Spokesmen and women

It is important to scrutinise the exact settings in which encounters between authors and dissenters take place. When we disbelieve the scientific literature, we are led from the many libraries around to the *very few* places where this literature is produced. Here we are welcomed by the author who shows us where the figure in the text comes from. Once presented with the instruments, who does the talking during these visits? At first, the authors: they *tell* the visitor what to *see*: 'see the endorphin effect?', 'look at the neutrinos!' However, the authors are not lecturing the visitor. The visitors have their faces turned towards the instrument and are watching the place where the thing is writing itself down (inscription in the form of collection of specimens, graphs, photographs, maps – you name it). When the dissenter was reading the scientific text it was difficult for him or her to doubt, but with imagination, shrewdness and downright awkwardness it was always possible. Once in the lab, it is much more difficult because the dissenters see with their own eyes. If we leave aside the many other ways to avoid going through the laboratory that we will study later, the dissenter does not have to believe the paper nor even the scientist's word since in a self-effacing gesture the author has stepped aside. 'See for yourself' the scientist says with a subdued and maybe ironic smile. 'Are you convinced now?' Faced with the thing itself that the technical paper was alluding to, the dissenters now have a choice between either accepting the fact or doubting their own sanity – the latter is much more painful.

We now seem to have reached the end of all possible controversies since there is nothing left for the dissenter to dispute. He or she is right in front of the thing he or she is asked to believe. There is almost no human intermediary between thing and person; the dissenter is in the very place where the thing is said to happen and at the very moment when it happens. When such a point is reached it seems that there is no further need to talk of 'confidence': the thing impresses itself directly on us. Undoubtedly, controversies are settled once and for all when such a situation is set up – which again is very rarely the case. The dissenter becomes a believer, goes out of the lab, borrowing the author's claim and confessing that 'X has incontrovertibly shown that A is B'. A new fact has been made which will be used to modify the outcome of some other controversies (see Part B, Section 3).

If this were enough to settle the debate, it would be the end of this book. But . . . there is someone saying 'but, wait a minute . . .' and the controversy resumes!

What was imprinted on us when we were watching the guinea pig ileum assay? 'Endorphin of course,' the Professor *said*. But what did we *see*? This

(3)

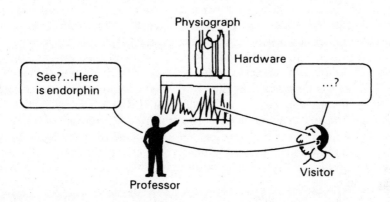

Figure 2.3

With a minimum of training we see peaks; we gather there is a base line, and we see a depression in relation to one coordinate that we understand to indicate the time. This is not endorphin yet. The same thing occurred when we paid a visit to Davis's gold and neutrino mine in South Dakota. We saw, he said, neutrinos counted straight out of the huge tank capturing them from the sun. But what *did* we see? Splurges on paper representing clicks from a Geiger counter. Not neutrinos, yet.

When we are confronted with the instrument, we are attending an 'audio-visual' spectacle. There is a *visual* set of inscriptions produced by the instrument and a *verbal* commentary uttered by the scientist. We get both together. The effect on conviction is striking, but its cause is mixed because we cannot differentiate what is coming from the thing inscribed, and what is coming from the author. To be sure, the scientist is not trying to influence us. He or she is simply commenting, underlining, pointing out, dotting the i's and crossing the t's, not adding anything. But it is also certain that the graphs and the clicks by themselves would not have been enough to form the image of endorphin coming out of the brain or neutrinos coming out of the sun. Is this not a strange situation? The scientists do not say anything more than what is inscribed, but without their commentaries the inscriptions say considerably less! There is a word to describe this strange situation, a very important word for everything that follows, that is the word **spokesman** (or **spokeswoman**, or **spokesperson**, or mouthpiece). The author behaves as if he or she were the mouthpiece of what is inscribed on the window of the instrument.

The spokesperson is someone who speaks for others who, or which, do not speak. For instance a shop steward is a spokesman. If the workers were gathered together and they all spoke at the same time there would be a jarring cacophony. No more meaning could be retrieved from the tumult than if they had remained silent. This is why they designate (or are given) a delegate who speaks on their behalf, and in their name. The delegate – let us call him Bill – does not speak in *his* name and when confronted with the manager does not speak 'as Bill' but as the 'workers' voice'. So Bill's longing for a new Japanese car or his note to get a pizza for his old mother on his way home are not the right topics for the meeting. The voice of the floor, articulated by Bill, wants a '3 per cent pay rise—and they are deadly serious about it, sir, they are ready to strike for it,' he tells the manager. The manager has his doubts: 'Is this really what they want? Are they really so adamant?' 'If you do not believe me,' replies Bill, 'I'll show you, but don't ask for a quick settlement. I told you they are ready to strike and you will see more than you want!' What does the manager see? He does not see what Bill said. Through the office window he simply sees an assembled crowd gathered in the aisles. Maybe it is because of Bill's interpretation that he reads anger and determination on their faces.

For everything that follows, it is very important not to limit this notion of spokesperson and not to impose any clear distinction between 'things' and 'people' in advance. Bill, for instance, represents people who could talk, but who, in fact, cannot all talk at once. Davis represents neutrinos that cannot talk, in principle, but which are made to write, scribble and sign thanks to the device set up by Davis. So in practice, there is not much difference between people and things: they both need someone to talk for them. From the spokesperson's point of view there is thus no distinction to be made between representing people and representing things. In each case the spokesperson literally does the talking for who or what cannot talk. The Professor in the laboratory speaks for endorphin like Davis for the neutrinos and Bill for the shopfloor. In our definition the crucial element is not the quality of the represented but only their number and the unity of the representative. The point is that confronting a spokesperson is not like confronting any average man or woman. You are confronted not with Bill or the Professor, but with Bill and the Professor *plus* the many things or people on behalf of whom they are talking. You do not address Mr Anybody or Mr Nobody but Mr or Messrs Manybodies. As we saw in the chapter on literature, it may be easy to doubt one person's word. Doubting a spokesperson's word requires a much more strenuous effort however because it is now one person – the dissenter – against a crowd – the author.

On the other hand, the strength of a spokesperson is not so great since he or she is by definition *one* man or woman whose word could be dismissed – one Bill, one Professor, one Davis. The strength comes from the representatives' word when they do not talk by and for themselves but *in the presence of* what they represent. Then, and only then, the dissenter is confronted simultaneously with the spokespersons and what they speak for: the Professor and the endorphin made visible in the guinea pig assay; Bill and the assembled workers; Davis and his solar neutrinos. The solidity of what the representative says is directly supported by the silent but eloquent presence of the represented. The result of such a set-up is that it seems as though the mouthpiece does not 'really talk', but that he or she is just commenting on what you yourself directly see, 'simply' providing you with the words you would have used anyway.

This situation, however, is the source of a major weakness. Who is speaking? The things or the people *through* the representative's voice? What does she (or he, or they, or it) say? Only what the things they represent would say if they could talk directly. But the point is that they cannot. So what the dissenter sees is, in practice, rather different from what the speaker says. Bill, for instance, says his workers want to strike, but this might be Bill's own desire or a union decision relayed by him. The manager looking through the window may see a crowd of assembled workers who are just passing the time and can be dispersed at the smallest threat. At any rate do they really want 3 per cent and not 4 per cent or 2 per cent? And even so, is it not possible to offer Bill this Japanese car he so dearly wants? Is the 'voice of the worker' not going to change his/its mind if the manager offers a new car to Bill? Take endorphin as another instance. What we really saw was a tiny depression in the regular spikes forming the base line. Is this the same as the one triggered by morphine? Yes it is, but what does that prove? It may be that all sorts of chemicals give the same shape in this peculiar assay. Or maybe the Professor so dearly wishes his substance to be morphine-like that he unwittingly confused two syringes and injected the same morphine twice, thus producing two shapes that indeed look identical.

What is happening? The controversy flares even after the spokesperson has spoken and displayed to the dissenter what he or she was talking about. How can the debate be stopped from proliferating again in all directions? How can all the strength that a spokesman musters be retrieved? The answer is easy: by letting the things and persons represented *say for themselves the same thing that the representatives claimed they wanted to say*. Of course, this never happens since they are designated because, by definition, such direct communication is impossible. Such a situation however may be convincingly staged.

Bill is not believed by the manager, so he leaves the office, climbs onto a podium, seizes a loudspeaker and asks the crowd, 'Do you want the 3 per cent rise?' A roaring 'Yes, our 3 per cent! Our 3 per cent!' deafens the manager's ears even through the window pane of his office. 'Hear them?' asks Bill with a modest but triumphant tone when they are sitting down again at the negotiating table. Since the workers themselves said exactly what the 'workers' voice' had said, the manager cannot dissociate Bill from those he represents and is really confronted with a crowd acting as one single man.

The same is true for the endorphin assay when the dissenter, losing his temper, accuses the Professor of fabricating facts. 'Do it yourself,' the Professor says, irritated but eager to play fair. 'Take the syringe and see for yourself what the assay reaction will be .' The visitor accepts the challenge, carefully checks the labels on the two vials and first injects morphine into the tiny glass chamber. Sure enough, a few seconds later the spikes start decreasing and after a minute or so they return to the base line. With the vial labelled endorphin, the very same result is achieved with the same timing. A unanimous, incontrovertible answer is thus obtained by the dissenter himself. What the Professor said the endorphin assay will answer, if asked directly, is answered by the assay. The Professor cannot be dissociated from his claims. So the visitor has to go back to the 'negotiating table' confronted not with the Professor's own wishes but with a Professor simply transmitting what endorphin really is.

No matter how many resources the scientific paper might mobilise, they carry little weight compared with this rare demonstration of power: the author of the claim steps aside and the doubter sees, hears and touches the inscribed things or the assembled people that reveal to him or to her exactly the same claim as the author.

(3) Trials of strength

For us who are simply following scientists at work there is no exit from such a set-up, no back door through which to escape the incontrovertible evidence. We have already exhausted all sources of dissent; indeed we might have no energy left to maintain the mere idea that controversy might still be open. For us laymen, the file is now closed. Surely, the dissenter we have shadowed since the beginning of Chapter 1 will give up. If the things say the same as the scientist, who can deny the claim any longer? How can you go any further?

The dissenter goes on, however, with more tenacity than the laymen. The identical tenor of the representative's words and the answers provided by the represented were the result of a carefully staged situation. The instruments needed to be working and finely tuned, the questions to be asked at the right time and in the right format. What would happen, asks the dissenter, if we stayed longer than the show and went backstage; or were to alter any of the many elements which, everyone agrees, are necessary to make up the whole instrument? The unanimity between represented and constituency is like what an inspector sees of a hospital or of a prison camp when his inspection is announced in advance. What if he steps outside his itinerary and tests the solid ties that link the represented and their spokesmen?

The manager, for instance, heard the roaring applause that Bill received, but he later obtains the foremen's opinion: 'The men are not for the strike at all, they would settle for 2 per cent. It is a union order; they applauded Bill because that's the way to behave on the shopfloor, but distribute a few pay rises and lay off a few ringleaders and they will sing an altogether different song.' In place of the unanimous answer given by the assembled workers, the manager is now faced with an *aggregate* of possible answers. He is now aware that the answer he got earlier through Bill was extracted from a complex setting which was at first invisible. He also realises that there is room for action and that each worker may be made to behave differently if pressures other than Bill's are exerted on them.

The next time Bill screams 'You want the 3 per cent don't you?' only a few half-hearted calls of agreement will interrupt a deafening silence.

Let us take another example, this time from the history of science. At the turn of the century, Blondlot, a physicist from Nancy, in France, made a major discovery like that of X-rays.[3] Out of devotion to his city he called them 'N-rays'. For a few years, N-rays had all sorts of theoretical developments and many practical applications, curing diseases and putting Nancy on the map of international science. A dissenter from the United States, Robert W. Wood, did not believe Blondlot's papers even though they were published in reputable journals, and decided to visit the laboratory. For a time Wood was confronted with incontrovertible evidence in the laboratory at Nancy. Blondlot stepped aside and let the N-rays inscribe themselves straight onto a screen in front of Wood. This, however, was not enough to get rid of Wood, who obstinately stayed in the lab asking for more experiments and himself manipulating the N-ray detector. At one point he even surreptitiously removed the aluminium prism which was generating the N-rays. To his surprise, Blondlot on the other side of the dimly lit room kept obtaining the same result on his screen even though what was deemed the most crucial element had been removed. The direct signatures made by the N-rays on the screen were thus made by something else. The unanimous support became a cacophony of dissent. By removing the prism, Wood severed the solid links that attached Blondlot to the N-rays. Wood's interpretation was that Blondlot so much wished to discover rays (at a time when almost every lab in Europe was christening new rays) that he unwittingly made up not only the N-rays, but also the instrument to inscribe them. Like the manager above, Wood realised that the coherent whole he was presented with was an aggregate of many elements that could be induced to go in many different directions. After Wood's action (and that of other dissenters) no one 'saw' N-rays any more but only smudges on photographic plates when Blondlot presented his N-rays. Instead of enquiring about the place of N-rays in physics, people started enquiring about the role of auto-suggestion in experimentation! The new fact had been turned into an artefact. Instead of going down the ladder of Figure 1.9, it went up the ladder and vanished from view.

The way out, for the dissenter, is not only to dissociate and disaggregate the many supporters the technical papers were able to muster. It is also to shake up the complicated set-up that provides graphs and traces in the author's laboratory in order to see how resistant the array is which has been mobilised in order to convince everyone. The work of disbelieving the literature has now been turned into the difficult job of manipulating the hardware. We have now reached another stage in the escalation between the author of a claim and the disbeliever, one that leads them further and further into the details of what makes up the inscriptions used in technical literature.

Let us continue the question-and-answer session staged above between the Professor and the dissenter. The visitor was asked to inject morphine and endorphin himself in order to check that there was no foul play. But the visitor is now more devious and does not make any effort to be polite. He wants to check where the vial labelled endorphin comes from. The Professor, unruffled, shows him the protocol book with the same code number as on the vial, a code that corresponds to a purified sample of brain extract. But this is a text, another piece of literature, simply an account book that could have been either falsified or accidentally mislabelled.

By now, we have to imagine a dissenter boorish enough to behave like a police inspector suspecting everyone and believing no one and finally wanting to see the real endorphin with his own eyes. He then asks, 'Where do I go from this label in the book to where the contents of the vial comes from?' Exasperated, the author leads him towards another part of the laboratory and into a small room occupied by glass columns of various sizes, filled with a white substance, through which a liquid is slowly percolating. Underneath the columns, a small piece of apparatus moves a rack of tiny flasks in which the percolated liquid is collected every few minutes. The continous flow at the top of the columns is collected, at the bottom, into a discrete set of flasks, each of which contains the part of the liquid that took the same given amount of time to travel through the column.

(4)–Here it is, says the guide, here is your endorphin.

–Are you kidding, replies the dissenter, where is endorphin? I don't see a thing?

–Hypothalamic brain extract is deposited on the top of the Sephadex column. It is a soup. Depending on what we fill it with, the column disassociates the mixture, sieves it; it may be done by gravity, or electrical charge, anything. At the end you get racks that collect samples which have behaved similarly in the column. This is called a fraction collector. Each fraction is then checked for purity. *Your* vial of endorphin came from *this* rack two days ago, no. 23/16/456.

–And this is what you call pure? How do I know it is pure? Maybe there are hundreds of brain extracts that travel through the column at the same pace exactly and end up in the same fraction.

The pressure is mounting. Everyone in the lab is expecting an outburst of rage, but the Professor politely leads the visitor towards another part of the laboratory.

(5)–Here is our new High Pressure Liquid Chromatograph (HPLC). See these tiny columns? They are like the ones you just saw, but each fraction collected there is submitted to an enormous pressure here. The column delays the passage and at this pressure it strongly differentiates the molecules. The ones that arrive at the same time at the end are *the same* molecules, the same, my dear colleague. Each fraction is read through an optical device that measures its optical spectrum. Here is the chart that you get . . .. See? Now, when you get a single peak it means the material is pure, so pure that a substance with only one different amino-acid in a hundred will give you *another* peak. Is not that quite convincing?

–(silence from the dissenter)

–Oh, I know! Maybe you are uncertain that I did the experiment with *your* vial of endorphin? Look here in the HPLC book. Same code, same time. Maybe you claim that I asked this gentleman here to fake the books, and obtain this peak for me with another substance? Or maybe you doubt the measurement of optical spectra. Maybe you think it is an obsolete piece of physics. No such luck, my dear colleague, Newton described this phenomenon quite accurately – but maybe he's not good enough for you.

The Professor's voice is quivering with hardly suppressed rage but he still behaves. Of course the dissenter could start doubting the HPLC or the fraction collector as he did with the guinea pig ileum assay, converting them from black boxes into a field of contention. He *could* in principle, but he *cannot* in practice since time is running out and he is sensitive to the exasperation in everyone's voice. And who is he anyway to mount a dispute against Water Associates, the company who devised this HPLC prototype? Is he ready to cast doubt on a result that has been accepted unquestioningly for the past 300 years, one that has been embedded in thousands of contemporary instruments? What he wants is to see endorphin. The rest, he must face it, cannot be disputed. He has to compromise and to admit that the Sephadex column, and the HPLC, are indisputable. In a conciliatory tone he says:

(6)–This is very impressive; however I must confess a slight disappointment. What I see here is a peak which, I admit, means that the brain extract is now pure. But how do I know that this pure substance is endorphin?

With a sigh, the visitor is led back to the assay room where the little guinea pig gut is still regularly contracting.

(7)–Each of the fractions deemed pure by the HPLC is tried out here, in this assay. Of all the pure fractions only two display any activity, I repeat only two. When the whole process is repeated in order to get purer material, this activity dramatically increases. The shape may be exactly superimposed onto that of commercially available morphine. Is that insignificant? We did it thirty-two times! Is that nothing? Each modification of the spikes has been tested for statistical significance. Only endorphin and morphine have any significant effect. Does all of that count for nothing? If you are so clever, can you give me an alternative explanation why morphine and this pure substance X would behave identically? Can you even imagine another explanation?
–No, I must admit, whispers the believer, I am very impressed. This really looks like genuine endorphin. Thank you so much for the visit. Don't trouble yourselves, I will find my own way out . . .. (exit the dissenter)

This exit is not the same as that of the semiotic character of Chapter 1, p.53. This time it is for good. The dissenter tried to disassociate the Professor from his endorphin, and he failed. Why did he fail? Because the endorphin constructed in the Professor's lab *resisted* all his efforts at modification. Every time the visitor followed a lead he reached a point where he had either to quit or start a new controversy about a still older and more generally accepted fact. The Professor's claim was tied to the brain, to the HPLC, to the guinea pig ileum assay. There is something in his claim that is connected to classic claims in physiology, pharmacology, peptide chemistry, optics, etc. This means that when the doubter tries out the connections, all these other facts, sciences and black boxes come to the Professor's rescue. The dissenter, if he doubts endorphin, has also to doubt Sephadex columns, HPLC technics, gut physiology, the Professor's honesty, that of his whole lab, etc. Although 'enough is never enough' – see the introduction – there is a point where no matter how pig-headed the dissenter could be, enough is enough. The dissenter would need so much more time, so many more allies and resources to continue to dissent that he has to quit, accepting the Professor's claim as an established fact.

Wood, who did not believe in N-rays, also tried to shake the connection between Blondlot and his rays. Unlike the former dissenter he succeeded. To dislocate the black boxes assembled by Blondlot, Wood did not have to confront the whole of physics, only the whole of one laboratory. The manager who suspected the workers' determination tried out the connections between them and their union boss. These connections did not resist a few classic clever tricks for long. In the three cases the dissenters imposed a showdown running from the claim to what supports the claim. When imposing such a **trial of strength** they are faced with spokespersons and what (or whom) these persons speak for. In some cases the dissenters isolate the representative from his or her 'constituency', so to speak; in other cases such a separation is impossible to obtain. It cannot be obtained without a trial of strength, any more than a boxer can claim to be a world champion without convincingly defeating the previous world champion. When the dissenter succeeds, the spokesperson is transformed from someone who speaks for others into someone who speaks for him or herself, who represents only him or herself, his or her wishes and fancies. When the dissenter fails, the spokesperson is seen not really as an individual but as the mouthpiece of many other mute phenomena. Depending on the trials of strength, spokespersons are turned into **subjective** individuals or into **objective** representatives. Being objective means that no matter how great the efforts of the disbelievers to sever the links between you and what you speak for, the links resist. Being subjective means that when you talk *in the name* of people or things, the listeners understand that you represent only yourself. From Mr Manybodies you are back to being Mr. Anybody.

It is crucial to grasp that these two adjectives ('objective', 'subjective') are *relative* to trials of strength in specific settings. They cannot be used to qualify a spokesperson or the things he or she is talking about once and for all. As we saw in Chapter 1, each dissenter tries to transform a statement from objective to subjective status, to transform, for instance, an interest in N-rays inside physics into an interest in self-suggestion in provincial laboratories. In the endorphin example, the dissenter seemed to be trying very hard to convert the Professor's claim into a subjective flight of fancy. In the end it was the lonely dissenter who saw his naive questioning turned into a trivial flight of fancy, if not an obsessive drive to seek fraud and find fault everywhere. In the trial of strength the Professor's endorphin was made *more objective* – going down the ladder – and the dissenter's counter-claim was made *more subjective* – pushed up the ladder. 'Objectivity' and 'subjectivity' are relative to trials of strength and they can shift gradually, moving from one to the other, much like the balance of power between two armies. A dissenter accused by the author of being subjective must now wage another struggle if he or she wishes to go on dissenting without being isolated, ridiculed and abandoned.

Part B
Building up counter-laboratories

Let me summarise our trip from the discussion at the beginning of Chapter 1 up to this point. What is behind the claims? Texts. And behind the texts? More texts, becoming more and more technical because they bring in more and more papers. Behind these articles? Graphs, inscriptions, labels, tables, maps, arrayed in tiers. Behind these inscriptions? Instruments, whatever their shape, age and cost that end up scribbling, registering and jotting down various traces. Behind the instruments? Mouthpieces of all sorts and manners commenting on the graphs and 'simply' saying what they mean. Behind them? Arrays of instruments. Behind those? Trials of strength to evaluate the resistance of the ties that link the representatives to what they speak for. It is not only words that are now lined up to confront the dissenter, not only graphs to support the words and references to support the whole assembly of allies, not only instruments to generate endless numbers of newer and clearer inscriptions, but, behind the instruments, new objects are lined up which are defined by their resistance to trials. Dissenters have now done all they can do to disbelieve, disaggregate and disassociate what is mustered behind the claim. They have come a long way since barging into the first discussion at the beginning of Chapter 1. They became readers of technical literature, then visitors to the few laboratories from which the papers were coming, then impolite inspectors manipulating the instruments to check how faithful they were to the author.

At this point they have to take another step – either give up, or find other resources to overcome the author's claim. In the second part of this book we will see that there exist many ways to reject the laboratory results (Chapter 4); but for this chapter we will concentrate on the rarest outcome, when, all else being equal, there is no other way open to the dissenters than to *build another laboratory*. The price of dissent increases dramatically and the number of people able to continue decreases accordingly. This price is entirely determined by the authors whose claims one wishes to dispute. The dissenters cannot do less than the authors. They have to gather more forces in order to untie what attaches the spokesmen and their claims. This is why all laboratories are *counter-laboratories* just as all technical articles are counter-articles. So the dissenters do not simply have to get a laboratory; they have to get a *better* laboratory. This makes the price still higher and the conditions to be met still more unusual.

(1) Borrowing more black boxes

How is it possible to obtain a better laboratory, that is a laboratory producing less disputable claims and allowing the dissenter – now head of a lab – to disagree and be believed? Remember what happened to the visitor to the Professor's laboratory. Every time a new flaw appeared which the disbeliever tried to exploit, the Professor presented him with a new and seemingly incontrovertible black box: a Sephadex column, an HPLC machine, basic physics, or classic physiology, etc. It might have been possible to dispute each of these, but it was not practical because the same energy would have been needed to reopen each of these black boxes. Indeed, *more* energy would have been applied because each of these facts in turn would have led to more tightly sealed black boxes: the microprocessors treating the data from the HPLC, the fabrication of the gel in the columns, the raising of guinea pigs in the animal quarters, the production of morphine at an Ely-Lily factory, etc. Each fact could be made the departure point of a new controversy that would have led to many more accepted facts, and so on *ad infinitum.*

The claim is tied to
too many blackboxes
for the dissenter
to untie them all

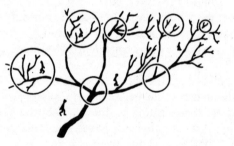

Figure 2.4

The dissenter was thus confronted by an exponential curve, a slope similar to the one drawn in Figure 1.8. Now that he has become the head of a brand new laboratory, one of the ways to make it a better counter-lab is to discover ways either of levelling the slope or of confronting his opponents with an even steeper one.

For instance Schally, in order to back up his ill-fated GHRH – see Chapter 1, statement (5) – used a bioassay called the rat tibia cartilage assay. Guillemin, who disagreed with GHRH, started to try out the tibia assay in exactly the way our dissenter tried out his guinea pig ileum assay.[4] In the face of this challenge, Schally's tibia assay was made to say quite different things by Guillemin. The growth of tibia cartilage in the rat might be caused by a growth hormone substance but might just *as well* have been caused by a variety of other chemicals, or indeed not have occurred at all. In several harsh papers, Guillemin said the 'results were so erratic that Schally's claims should be taken with the most extreme precaution'. Thus Schally was cut off from his supply line. He claimed the existence of GHRH, but nothing followed. Isolated, his claim was made more subjective by the dissenter's action.

Why should anyone believe Guillemin's counter-claim rather than Schally's claim? One obvious way to strengthen this belief is to modify the bioassay to make it impossible for anyone to make it say different things from Guillemin. Guillemin discarded the rat tibia assay and shifted to a rat pituitary cell culture. Instead of seeing the growth of cartilage with the naked eye, what is 'seen' is the amount of hormone released by the few pituitary cells maintained in a culture; this amount is measured by an instrument–in the sense I gave this term earlier–called radio-immunoassay. The new assay is much *more* complicated than Schally's older ones–in itself the radio-immunoassay requires several technicians and takes up to a week to complete–but it gives inscriptions at the end that may be said to be more clear-cut, that is they literally cut shapes out of the background. In other words, even without understanding a word of the issue, the perceptive judgment to be made on one is easier than on the other.

The answers are less equivocal than the 'erratic' ones given by the tibia assay–that is, they leave less room for the dissenter to quibble–and the whole instrument is *less* easily disputable. Although it is complicated, the cell culture assay can be taken as *a single black box* which provides a single window from which to read the amount of GHRH. Naturally, it can be disputed in principle. It is just that it's harder to do so in practice. A physiologist with a little training may nitpick at the cartilage assay, may quibble about the length from growth in the tibia. He or she needs much more than a little training to dispute Guillemin's new figures. The assay is now tied to basic advances in molecular biology, immunology and the physics of radioactivity. Nitpicking at the inscriptions is possible but less reasonable, the heckler needing more resources and becoming more isolated. The gain in conviction is clear: from Schally's first words a fierce dispute ensues about the assay which is supposed to reveal the very existence of GHRH. In Guillemin's counter-paper this part of the discussion *at least* has been sealed off since his detection system is made indisputable, and the range of possible disputes has *shifted to other* aspects of the same claims.

Another example is provided by the controversy about the detection of gravitational waves.[5] One physicist, Weber, built a massive antenna made of a large aluminium alloy bar weighing several tons that vibrated at a certain frequency. To detect a gravitational wave the antenna must be insulated from all other influences–ideally it should be in a vacuum, free from seismic vibrations and radio interference, at a temperature at or near absolute zero, etc. Taken as an instrument, the whole set-up provides a window which allows one to read the presence of gravitational waves. The problem is that the peaks above the noise threshold are so tiny that any passing physicist could dispute Weber's claim. Indeed, any passing physicist could set the instrument off! Weber argues that they represent gravitation but every dissenter may claim that they represent many other things *as well.* This little expression 'as well' is what kills most solid claims. As long as it is possible to say 'as well', there is no established line from the gravitation waves to Weber via the antenna. The figure offered by Weber may represent either 'gravitational waves' or meaningless scribbles registering terrestrial noise. To be sure, there are many ways out of the controversy so as to shrug off Weber's claim as a mere opinion. But the way out of the controversy that interests us here is to build *another* antenna, one, for instance, that is a thousand million times more sensitive than Weber's so that this part of the detection at least is not disputed. The aim of this new antenna is to confront the sceptic with an incontrovertible black box *earlier* in the process. After this, sceptics may still discuss the amount of gravitation, and what it does to the relativity theory or to astrophysics, but they will not argue that there are peaks that cannot be explained by terrestrial interferences. With the first antenna alone, Weber might be the freak and the dissenters the sensible professionals. With the new antenna, those who deny the presence of the peaks are the isolated sceptics and it is Weber who is the sensible professional. All other things being equal the balance of power would have been tipped. (In this case, however, it did not make the slightest difference because many other avenues for dissent were opened.)

Borrowing more black boxes and situating them earlier in the process is the first obvious strategy for building a better counter-laboratory. The discussion is diffracted and shunted away. Any one laboratory gets an edge on all the others if it finds a way to delay the possible discussions until later. In the early days of microbe cultures, for example, the microbes were grown in a liquid like urine. They were visible in the flasks but you needed keen and trained eyesight to detect them. Dissent could ensue because the construction of the fact was interrupted from the start by a preliminary discussion on whether or not microbes were present in the flask. When Koch invented the solid milieu culture, acute eyesight was no longer needed to see the little microbes: they made nice little coloured patches which contrasted clearly with the white background. The visibility was dramatically enhanced when specific dyes coloured certain microbes or their parts. The laboratory endowed with these techniques made dissent more difficult: a slope was deepened, a trench was dug. Although many other aspects were still open to dispute, the presence of the microbes was made indisputable.

At this point, it is easy to imagine the growing differences between good and bad (counter-) laboratories. Imagine a lab that starts making claims based on the cartilage tibia assay, Weber's first antenna and the liquid microbe culture. If the head of this laboratory wanted to be believed he would have an endless task. Every time he opened his mouth, any number of his dear colleagues would start shaking their heads, and suggesting many alternatives just as plausible as the first. To do so, they would only need a bit of imagination. Like Achilles in Zeno's paradox, the challenger will never reach the end of his argument since each point will be the start of an indefinite regression. In contrast, claims produced by the good laboratory cannot be opposed simply with a bit of imagination. The cost of disputing the claims increases proportionally with the number of black boxes assembled by the author. Faced with the pituitary culture assay, the new antenna which is one thousand million times more sensitive and the solid milieu culture, the dissenters are forced to assent or, at least, *to redirect* their dissent toward some other aspect of the claims. They can still mount a controversy but the magnitude of the mobilisation needed to do so has increased. They need an even better equipped laboratory with more and more black boxes, thus delaying the dispute still further. The vicious (or virtuous) circle of lab construction is now launched and there is no way to stop it – apart from giving up the production of credible arguments altogether, or recruiting more powerful allies elsewhere.

(2) Making actors betray their representatives

The competition between scientists – whom I will treat in this section as alternately authors and dissenters – to turn one another's claims into subjective opinion leads to expensive laboratories equipped with more and more black boxes introduced as early as possible into the discussion. This game, however, would soon stop if only existing black boxes were mobilised. After a time dissenters and authors – all things remaining equal – would have access to the *same* equipment, would tie their claims to the same harder, colder and older facts and none would be able to get an edge on the other: their claims would be thus left in limbo, in intermediary stages between fact and artefact, objectivity and subjectivity. The only way to break this stalemate is to find either new and unexpected resources (see the next section) or, more simply, to force the opponent's allies to *change camp*.

This would happen, for instance, if the manager of our little vignette above could organise a secret ballot to decide about the continuation of the strike. Remember that Bill, the shop steward, claimed that 'all the workers want a 3 per cent pay rise'. This claim was confirmed at meetings during which the represented said the same things as their mouthpiece. Even if the manager suspects that the workers are not so unanimous, each public meeting loudly confirms Bill's claim. However, in organising a secret ballot, the manager tests the same actors in a different way, by exerting a new set of pressures on them: isolation, secrecy, recounting of the ballots, surveillance. Submitted to these new trials, only 9 per cent of the same workers voted for the continuation of the strike, and 80 per cent were ready to settle for 2 per cent. The represented have changed camp. They now say what the manager said they would say. They have a new spokesperson. This, naturally, does not stop the controversy, but the dispute will now bear on the election process itself. Bill and his union accuse the manager of intimidation, unfair pressure, of having stuffed the ballot boxes and so on. This shows that even the most faithful supporters of a spokesman may be made to *betray*.

As I showed above, both people able to talk and things unable to talk have spokesmen (Part A, section 2). I propose to call whoever and whatever is represented **actant**. What the manager did to Bill, a dissenter may do for the ally of his opponent's laboratory. Pouchet, engaged in a bitter struggle against Louis Pasteur's claim that there is no spontaneous generation, built a nice counter-experiment.[6] Pasteur argued that it is always germs introduced from the outside that generate micro-organisms. Long swan-necked open glass flasks containing sterilised infusion were contaminated at low altitude but stayed sterile in the High Alps. This impressive series of demonstrations established an incontrovertible link between a new actor, the micro-organisms, and what Pasteur said they could do: microbes could not come from *within* the infusion but only from *outside*. Pouchet, who rejected Pasteur's conclusion, tried out the connection and forced the micro-organisms to emerge from within. Repeating Pasteur's experiment Pouchet showed that glass flasks containing a sterile hay infusion were very soon swarming with micro-organisms even in the 'germ-free' air of the Pyrenees Mountains. The micro-organisms on which Pasteur depended were made to betray him: they appeared spontaneously thus supporting Pouchet's position. In this case, the actants change camps and two spokesmen are supported at once. This change of camp does not stop the controversy, because it is possible to accuse Pouchet of having unknowingly introduced micro-organisms from outside even though he sterilised everything. The meaning of 'sterile' becomes ambiguous and has to be renegotiated. Pasteur, now in the role of dissenter, showed that the mercury used by Pouchet was contaminated. As a result Pouchet was cut off from his supply lines, betrayed by his spontaneous micro-organisms, and Pasteur becomes the triumphant spokesman, aligning 'his' micro-organisms which act on command. Pouchet failed in his dissent and ended up isolated, his 'spontaneous generation' reduced by Pasteur to a *subjective* idea, to be explained not by the behaviour of microbes but by the influence of 'ideology' and 'religion'.[7]

The same luring of allies away from their spokesperson occurred among the Samoans. As mobilised in the 1930s by Margaret Mead to act on North American ideals of education and sexual behaviour, Samoan girls were more liberated than Western ones and free from the crises of adolescence.[8] This well-established fact was attributed not to Mead – acting as the anthropologist mouthpiece of the Samoans – but to the Samoans. Recently another anthropologist, Derek Freeman, attacked Mead, severing all links between the Samoan girls and Margaret Mead. She was turned into an isolated liberal American lady without any serious contact with Samoa and writing a 'noble savage' fiction off the top of her head. Freeman, the new spokesman of the Samoans, said the girls there were sexually repressed, assaulted and often raped and that they went through a terrible adolescence. Naturally, this 'kidnapping', so to speak, of Samoan teenagers by a new representative does not bring the controversy to an end any more than in our other examples. The question is now to decide if Freeman is a boorish and insensitive male influenced by sociobiology, and if he has more Samoan allies on his side than Margaret Mead, a highly thought of female anthropologist, sensitive to all the subtle cues of her Samoan informants. The point for us is that the most sudden reversal in the trials of strength between authors and dissenters may be obtained simply by cutting the links tying them to their supporters.

A subtler strategy than Freeman's to cut these links was employed by Karl Pearson in his dispute with George Yule's statistics.[9] Yule had devised a coefficient to measure the strength of an association between two *discrete* variables. This crude but robust coefficient allowed him to decide whether or not there was an association between, for instance, vaccination and the death rate. Yule was not interested in defining links more precisely. All he wanted to be able to determine was whether vaccination decreased the death rate. Pearson, on the other hand, objected to Yule's coefficient because when you wanted to decide *how close* the links were, it offered a wide range of possible solutions. With Yule's coefficient you would never know, in Pearson's opinion, if you had your data all safely arrayed behind your claims. Yule did not bother because he was treating only discrete entities. Pearson, however, had a much more ambitious project and wanted to be able to mobilise a large number of *continuous* variables such as height, colour of skin, intelligence . . . With Yule's coefficient he would have been able to define only weak associations between genetic variables. This meant that any dissenter could easily have severed him from his data and turned one of the most impressive arrays on genetic determinism ever compiled into a mixed and disorderly crowd of unclear relations. Pearson devised a correlation coefficient which made any discrete variable the outcome of a continuous distribution. Yule was left with only weak associations and Pearson, tying his data together with his 'tetrachoric coefficient of correlation', could transform any continuous variable into a strongly associated whole of discrete variables and so *solidly* attach intelligence to heredity. This of course did not mark the end of the controversy. Yule tried out the Pearson coefficient showing that it arbitrarily transformed continuous variables into discrete ones. If successful, Yule would have deprived Pearson of the support of his data. Although this controversy has been continuing for nearly a hundred years, the lesson for us is that, with the same equipment and data, the stalemate between dissenting authors may be broken by a simple modification of what it is that ties the data together (we shall see more of this phenomenon in Chapter 6).

In each of the examples above I showed how allies were enticed away from their representative in order to tip the balance, but I also indicated that this need not settle the debate. Often it modifies the field of contention enough to buy time – not enough to win. This strategy must in general be combined with that of section 1 in order to succeed – borrowing more black boxes and positioning them earlier in the process – and with that of the third section, which is the most daring and the most difficult to grasp for the visiting layperson.

(3) Shaping up new allies

The dissenter, now the head of a (counter-) laboratory, has imported as many black-boxed instruments as possible and has tried to entice his opponent's supporters away. Even combining these two strategies he or she will not fare very well since all scientists are playing with a *limited set* of instruments and actants. After a few moves the controversy will reach a new stalemate with the supporters continually changing camp: for and against the manager, for and against Pasteur, for and against Margaret Mead, for and against Pearson, with no end in sight. No credible fact will be produced in such confusion since no third party will be able to borrow any statement as a black box to put it to use elsewhere. In order to break the stalemate, other allies which are *not yet* defined have to be brought in.

Let me go back to the example of GHRH discovered by Schally using his rat tibia cartilage assay. We saw how Guillemin, rejecting this 'discovery' – now in quotation marks – devised a new, less controvertible assay, the pituitary cell culture (Chapter 1, section 2). With it, he induced the GHRH supporting Schally's claim to shift alliances. Remember that when Schally thought he had found a new important hormone, Guillemin intervened and showed that this 'new important hormone' was a contaminant, a piece of haemoglobin. By following the two strategies we have just defined, Guillemin won but only *negatively*. Although he overcame his competitor, his own claims about GHRH – which he calls GRF – are not made more credible. For a third party the whole topic is simply a mess from which no credible fact emerges. In the search for the final *coup de grâce,* the dissenter needs something more, a supplement, a little 'je ne sais quoi' that, everything being equal, will ensure victory and convince the third party that the controversy has indeed been settled.

In the (counter-) laboratory the purified extracts of GRF are injected into the cell culture. The result is appalling: nothing happens. Worse than nothing, because the results are negative: instead of being triggered by GRF the growth hormone is decreased. Guillemin gives his collaborator, Paul Brazeau, who has done the experiment, a good dressing down.[10] The whole instrument, supposed to be a perfect black box, is called into doubt, and the whole career of Brazeau, supposed to be a skilled and honest worker, is jeopardised. The dissenter/author struggle has now shifted inside the laboratory and they are both trying out the assay, the purification scheme and the radio-immunoassay exactly as the visitor did above for endorphin (In Part A, section 3). At the third trial Brazeau still obtained the same result. That is, no matter how much effort he was making, the same negative results were produced. No matter how strongly Guillemin attacked him, he was led every time to the same sort of quandary with which I finished Part A: either to quit the game or to start discussing so many basic, old and accepted black boxes that the whole lab would have to be dismantled. Since the negative results resisted all trials of strength, since the cell culture assay was left indisputable, and since Brazeau's honesty and skill were withstanding the shock, some other weak point had to give way. The hormone they were looking for *released* growth hormone; in their hands it *decreased* growth hormone. Since they could no longer doubt that their 'hands' were good, they had to doubt the first definition or quit the game altogether: they had got their hands on a hormone that *decreased* the production of growth hormone. They had, in other words, tried out a *new* hormone, a new, unexpected and still undefined ally to support another claim. Within a few months they had obtained a decisive advantage over Schally. Not only had he confused GHRH with a piece of haemoglobin, but he had sought the wrong substance all along.

We have reached a point which is one of the most delicate of this book, because, by following dissenting scientists, we have access to their most decisive arguments, to their ultimate source of strength. Behind the texts, they have mobilised inscriptions, and sometimes huge and costly instruments to obtain these inscriptions. But something else resists the trials of strength behind the instruments, something that I will call provisionally a **new object**. To understand what this is, we should stick more carefully than ever to our method of following only scientists' practice, deaf to every other opinion, to tradition, to philosophers, and even to what scientists say about what they do (see why in the last part of this chapter).

What is a new object in the hands of a scientist? Consider the GRF that Guillemin and Brazeau were expecting to find: it was defined by its effect on tibia cartilage assay and in cell cultures. The effect was uncertain in the first assay, certain and negative in the second. The definition had to change. The new object, at the time of its inception, is still undefined. More exactly, it is defined by what it does in the laboratory trials, *nothing more, nothing less*: its tendency to decrease the release of growth hormone in the pituitary cells culture. The etymology of 'definition' will help us here since defining something means providing it with limits or edges (*finis*), giving it a shape. GRF had a shape; this shape was formed by the answers it gave to a series of trials inscribed on the window of an instrument. When the answers changed and could not be ignored a new shape was provided, a new thing emerged, a something, still unnamed, that did exactly the opposite of GRF. Observe that in the laboratory, the new object is *named after what it does*: 'something that inhibits the release of growth hormone'. Guillemin then invents a new word that summarises the actions defining the thing. He calls it 'somatostatin' – that which blocks the body (implying body growth).

Now that somatostatin is named and accepted, its properties have changed and are not of interest to us at this point. What counts for us is to understand the new object just at the moment of its emergence. Inside the laboratory the new object is *a list of written answers to trials*. Everyone today talks for instance of 'enzymes' which are well-known objects. When the strange things later called 'enzymes' were emerging among competing laboratories, scientists spoke of them in very different terms:[11]

(8) From the liquid produced by macerating malt, Payen and Persoz are learning to extract, through the action of alcohol, a solid, white, amorphous, neutral, more or

less tasteless substance that is insoluble in alcohol, soluble in water and weak
alcohol, and which cannot be precipitated by sub-lead acetate. Warmed from 65° to
75° with starch in the presence of water, it separates off a soluble substance, which is
dextrin.

At the time of its emergence, you cannot do better than explain what the new
object is by repeating the list of its constitutive actions: 'with A it does this, with C
it does that.' It has *no other shape than this list.* The proof is that if you add an item
to the list you *redefine the object*, that is, you give it a new shape. 'Somatostatin'
for instance was defined by the now well-established fact that, coming from the
hypothalamus, it inhibited the release of growth hormone. The discovery I
summarised above was described in this way for a few months after its
construction. When another laboratory added that somatostatin was also found
in the pancreas and inhibited not only growth hormone but also glucagon and
insulin production, the definition of somatostatin had to be changed, in the same
way as the definition of GRF had to be altered when Brazeau failed to get positive
results in his assay. The new object is completeley defined by the list of answers in
laboratory trials. To repeat this essential point in a lighter way, the new object is
always called after a name of actions summarising the trials it withstood like the
old Red Indian appellations 'Bear Killer' or 'Dread Nothing' or 'Stronger than a
Bison'!

In the strategies we have analysed so far, the spokesperson and the actants he
or she represented were already present, arrayed and well drilled. In this new
strategy the representatives are looking for actants they do not know and the only
thing they can say is to list the answers the actants make under trials.

Pierre and Marie Curie originally had no name for the 'substance x' they tried
out. In the laboratory of the Ecole de Chimie the only way to shape this new
object is to multiply the trials it undergoes, to attack it by all sorts of terrible
ordeals (acids, heat, cold, presure).[12] Will something resist all these trials and
tribulations? If so, then here it is, the new object. At the end of their long list of
'sufferings' undergone by the new substance (and also by the unfortunate Curies
attacked by the deadly rays so carelessly handled) the authors propose a new
name – 'polonium'. Today polonium is one of the radioactive elements; at the
time of its inception it was the long list of trials successfully withstood in the
Curies' laboratory:

> (9) Pierre and Marie Curie: –Here is the new substance emerging from this
> mixture, pitchblende, see? It makes the air become conductive. You can even
> measure its activity with the instrument that Pierre devised, a quartz electrometer,
> right here. This is how we follow our hero's fate through all his ordeals and
> tribulations.
> Scientific Objector: This is far from new, uranium and thorium are also active.
> –Yes, but when you attack the mixture with acids, you get a liquor. Then, when
> you treat this liquor with sulphurated hydrogen, uranium and thorium stay with the
> liquor, while our young hero is precipitated as a sulphuride.
> – What does that prove? Lead, bismuth, copper, arsenic and antimony all pass this

trial as well, they too are precipitated!

– But if you try to make all of them soluble in ammonium sulphate, the active something resists . . .

– Okay, I admit it is not arsenic, nor antimony, but it might be one of the well-known heroes of the past, lead, copper or bismuth.

– Impossible, dear, since lead is precipitated by sulphuric acid while the substance stays in solution; as for copper, ammoniac precipitates it.

– So what? This means that your so-called 'active substance' is simply bismuth. It adds a property to good old bismuth, that of activity. It does not define a new substance.

– It does not? Well, tell us what will make you accept that there is a substance?

– Simply show me one trial in which bismuth reacts differently from your 'hero'.

– Try heating it in a Boheme tube, under vacuum, at 700° centigrade. And what happens? Bismuth stays in the hottest area of the tube, while a strange black soot gathers in the cooler areas. This is more active than the material with which we started. And you know what? If you do this several times, the 'something' that you confuse with bismuth ends up being four hundred times more active than uranium!

– . . . .

– Ah, you remain silent . . . . We therefore believe that the substance we have extracted from pitchblende is a hitherto unknown metal. If the existence of this new metal is confirmed we propose to name it polonium after Marie's native country.

What are these famous things which are said to be behind the texts made of? They are made of a list of victories: it defeated uranium and thorium at the sulphurated hydrogen game; it defeated antimony and arsenic at the ammonium sulphur game; and then it forced lead and copper to throw in the sponge, only bismuth went all the way to the semi-final, but it too got beaten down during the final game of heat and cold! At the beginning of its definition the 'thing' is a *score list* for a series of trials. Some of these trials are imposed on it either by the scientific objector and tradition – for instance to define what is a metal – or tailored by the authors – like the trial by heat. The 'things' behind the scientific texts are thus similar to the heroes of the stories we saw at the end of Chapter 1: they are all defined by their **performances**. Some in fairy tales defeat the ugliest seven-headed dragons or against all odds they save the king's daughter; others inside laboratories resist precipitation or they triumph over bismuth . . . . At first, there is no other way to know the essence of the hero. This does not last long however, because each performance presupposes a **competence**[13] which retrospectively explains why the hero withstood all the ordeals. The hero is no longer a score list of actions; he, she or it is an essence slowly unveiled through each of his, her or its manifestations.

It is clear by now to the reader why I introduced the word 'actant' earlier to describe what the spokesperson represents. Behind the texts, behind the instruments, inside the laboratory, we do not have Nature – not yet, the reader will have to wait for the next part. What we have is an array allowing new extreme constraints to be imposed on 'something'. This 'something' is progressively shaped by its re-actions to these conditions. This is what is behind all the arguments we have analysed so far. What was the endorphin tried out by the dissenter in Part A, section 3? The superimposition of the traces obtained by: a sacrificed guinea pig whose gut was then hooked up to electric wires and regularly stimulated; a hypothalamus soup extracted after many trials from slaughtered sheep and then forced through HPLC columns under a very high pressure.

Endorphin, before being named and for as long as it is a new object, *is* this list *readable* on the instruments *in* the Professor's laboratory. So is a microbe long before being called such. At first it is something that transforms sugar into alcohol in Pasteur's lab. This something is narrowed down by the multiplication of feats it is asked to do. Fermentation still occurs in the absence of air but stops when air is reintroduced. This exploit defines a new hero that is killed by air but breaks down sugar in its absence, a hero that will be called, like the Indians above, 'Anaerobic' or 'Survivor in the Absence of Air'. Laboratories generate so many new objects because they are able to create extreme conditions and because each of these actions is obsessively inscribed.

This naming after what the new object does is in no way limited to actants like hormones or radioactive substances, that is to the laboratories of what are often called 'experimental sciences'. Mathematics also defines its subjects by what they *do*. When Cantor, the German mathematician, gave a shape to his transfinite numbers, the shape of his new objects was obtained by having them undergo the simplest and most radical trial:[14] is it possible to establish a one-to-one connection between, for instance, the set of points comprising a unit square and the set of real numbers between 0 and 1? It seems absurd at first since it would mean that there are as many numbers on one side of a square as in the whole square. The trial is devised so as to see if two different numbers in the square have different images on the side or not (thus forming a one-to-one correspondence) or if they have only one image (thus forming a two-to-one correspondence). The written answer on the white sheet of paper is incredible: 'I see it but I don't believe it,' wrote Cantor to Dedekind. There are as many numbers on the side as in the square. Cantor creates his transfinites from their performance in these extreme, scarcely conceivable conditions.

The act of defining a new object by the answers it inscribes on the window of an instrument provides scientists and engineers with their final source of strength. It constitutes our **second basic principle**, as important as the first in order to understand science in the making: scientists and engineers speak in the name of new allies that they have shaped and enrolled; representatives among other representatives, they add these unexpected resources to tip the balance of force in their favour. Guillemin now speaks for endorphin and somatostatin, Pasteur for visible microbes, the Curies for polonium, Payen and Persoz for enzymes, Cantor for transfinites. When they are challenged, they cannot be isolated, but on the contrary their constituency stands behind them arrayed in tiers and ready to say the same thing.

(4) Laboratories against laboratories

Our good friend, the dissenter, has now come a long way. He or she is no longer the shy listener to a technical lecture, the timid onlooker of a scientific experiment, the polite contradictor. He or she is now the head of a powerful laboratory utilising all available instruments, forcing the phenomena supporting the competitors to support him or her instead, and shaping all sorts of unexpected objects by imposing harsher and longer trials. The power of this laboratory is measured by the extreme conditions it is able to create: huge accelerators of millions of electron volts; temperatures approaching absolute zero; arrays of radio-telescopes spanning kilometres; furnaces heating up to thousands of degrees; pressures exerted at thousands of atmospheres; animal quarters with thousands of rats or guinea pigs; gigantic number crunchers able to do thousands of operations per millisecond. Each modification of these conditions allows the dissenter to mobilise one more actant. A change from micro to phentogram, from million to billion electron volts; lenses going from metres to tens of metres; tests going from hundreds to thousands of animals; and the shape of a new actant is thus redefined. All else being equal, the power of the laboratory is thus proportionate to the number of actants it can mobilise on its behalf. At this point, statements are not borrowed, transformed or disputed by empty-handed laypeople, but by scientists with whole laboratories *behind* them.

However, to gain the final edge on the opposing laboratory, the dissenter must carry out a fourth strategy: he or she must be able to transform the new objects into, so to speak, older objects and feed them back into his or her lab.

What makes a laboratory difficult to understand is not what is presently going on in it, but what *has been* going on in it and in other labs. Especially difficult to grasp is the way in which new objects are immediately transformed into something else. As long as somatostatin, polonium, transfinite numbers, or anaerobic microbes are shaped by the list of trials I summarised above, it is easy to relate to them: tell me what you go through and I will tell you what you are. This situation, however, does not last. New objects become **things**: 'somatostatin', 'polonium', 'anaerobic microbes', 'transfinite numbers', 'double helix' or '*Eagle* computers', things isolated from the laboratory conditions that shaped them, things with a name that now seem independent from the trials in which they proved their mettle. This process of transformation is a very common one and occurs constantly both for laypeople and for the scientist. All biologists now take 'protein' for an object; they do not remember the time, in the 1920s, when protein was a whitish stuff that was separated by a new ultracentrifuge in Svedberg's laboratory.[15] At the time protein was nothing but the action of differentiating cell contents by a centrifuge. Routine use however transforms the naming of an actant after what it does into a common name. This process is not mysterious or special to science. It is the same with the can opener we routinely use in our kitchen. We consider the opener and the skill to handle it as one black box which means that it is unproblematic and does not require planning and attention. We forget the many trials we had to go through (blood, scars, spilled beans and ravioli, shouting parent) before we handled it properly, anticipating the weight of the can, the reactions of the opener, the resistance of the tin. It is only when watching our own kids still learning it the hard way that we might remember how it was when the can opener was a 'new object' for us, defined by a list of trials so long that it could delay dinner for ever.

This process of routinisation is common enough. What is less common is the way the same people who constantly generate new objects to win in a controversy are also constantly transforming them into relatively older ones in order to win still faster and irreversibly. As soon as somatostatin has taken shape, a new bioassay is devised in which sosmatostatin takes the role of a stable, unproblematic substance in a trial set up for tracking down a new problematic substance, GRF. As soon as Svedberg has defined protein, the ultracentrifuge is made a routine tool of the laboratory bench and is employed to define the constituents of proteins. No sooner has polonium emerged from what it did in the list of ordeals above than it is turned into one of the well-known radioactive elements with which one can design an experiment to isolate a new radioactive substance further down in Mendeleev's table. The list of trials becomes a thing; it is literally *reified*.

This process of reification is visible when going from new objects to older ones, but it is also reversible although less visible when going from younger to older ones. All the new objects we analysed in the section above were framed and defined by stable black boxes which had *earlier* been new objects before being similarly reified. Endorphin was made visible in part because the ileum was known to go on pulsating long after guinea pigs are sacrificed: what was a new object several decades earlier in physiology was one of the black boxes participating in the endorphin assay, as was morphine itself. How could the new unknown substance have been compared if morphine had not been known? Morphine, which had been a new object defined by its trials in Seguin's laboratory sometime in 1804, was used by Guillemin in conjunction with the guinea pig ileum to set up the conditions defining endorphin. This also applies to the physiograph, invented by the French physiologist Marey at the end of the nineteenth century. Without it, the transformation of gut pulsation would not have been made graphically visible. Similarly for the electronic hardware that enhanced the signals and made them strong enough to activate the physiograph stylus. Decades of advanced electronics during which many new phenomena had been devised were mobilised here by Guillemin to make up another part of the assay for endorphin. Any new object is thus shaped by simultaneously importing many older ones in their reified form. Some of the imported objects are from young or old disciplines or pertain to harder or softer ones. The point is that the new object emerges from a complex set-up of sedimented elements each of which has been a new object at some point in time and space. The genealogy and the archaeology of this sedimented past is always possible in theory but becomes more and more difficult as time goes by and the number of elements mustered increases.

It is just as difficult to go back to the time of their emergence *as it is to contest them.* The reader will have certainly noticed that we have gone full circle from the first section of this part (borrowing more black boxes) to this section (blackboxing more objects). It is indeed a circle with a feedback mechanism that creates better and better laboratories by bringing in as many new objects as possible in as reified a form as possible. If the dissenter quickly re-imports somatostatin, endorphin, polonium, transfinite numbers as so many incontrovertible black boxes, his or her opponent will be made all the weaker. His or her ability to dispute will be decreased since he or she will now be faced with piles of black boxes, obliged to untie the links between more and more elements coming from a more and more remote past, from harder disciplines, and presented in a more reified form. Has the shift been noticed? It is now the author who is weaker and the dissenter stronger. The author must now either build a better laboratory in order to dispute the dissenter's claim and tip the balance of power back again, or quit the game – or apply one of the many tactics to escape the problem altogether that we will see in the second part of this book. The endless spiral has travelled one more loop. Laboratories grow because of the number of elements fed back into them, and this growth is irreversible since no dissenter/author is able to enter into the fray later with fewer resources at his or her disposal – everything else being equal. Beginning with a few cheap elements borrowed from common practice, laboratories end up after several cycles of contest with costly and enormously complex set-ups very remote from common practice.

The difficulty of grasping what goes on inside their walls thus comes from the sediment of what has been going on in other laboratories earlier in time and elsewhere in space. The trials currently being undergone by the new object they give shape to are probably easy to explain to the layperson – and we are all laypeople so far as disciplines other than our own are concerned – but the older objects capitalised in the many instruments are not. The layman is awed by the laboratory set-up, and rightly so. There are not many places under the sun where so many and such hard resources are gathered in so great numbers, sedimented in so many layers, capitalised on such a large scale. When confronted earlier by the technical literature we could brush it aside; confronted by laboratories we are simply and literally impressed. We are left without power, that is, without resource to contest, to reopen the black boxes, to generate new objects, to dispute the spokesmen's authority.

Laboratories are now powerful enough to define **reality**. To make sure that our travel through technoscience is not stifled by complicated definitions of reality, we need a simple and sturdy one able to withstand the journey: reality as the latin word *res* indicates, is what *resists*. What does it resist? *Trials of strength.* If, in a given situation, no dissenter is able to modify the shape of a new object, then that's it, it *is* reality, at least for as long as the trials of strength are not modified. In the examples above so many resources have been mobilised in the last two chapters by the dissenters to support these claims that, we must admit, resistance will be vain: the claim has to be true. The minute the contest stops, the minute I write the word 'true', a new, formidable ally suddenly appears in the winner's camp, an ally invisible until then, but behaving now as if it had been there all along: Nature.

Part C
Appealing (to) Nature

Some readers will think that it is about time I talked of Nature and the real objects *behind* the texts and behind the labs. But it is not I who am late in finally talking about reality. Rather, it is Nature who always arrives late, too late to explain the rhetoric of scientific texts and the building of laboratories. This belated, sometimes faithful and sometimes fickle ally has complicated the study of technoscience until now so much that we need to understand it if we wish to continue our travel through the construction of facts and artefacts.

(1) 'Natur mit uns'

'Belated?' 'Fickle?' I can hear the scientists I have shadowed so far becoming incensed by what I have just written. 'All this is ludicrous because the reading and the writing, the style and the black boxes, the laboratory set-ups – indeed all existing phenomena – are simply *means* to express something, vehicles for conveying this formidable ally. We might accept these ideas of 'inscriptions', your emphasis on controversies, and also perhaps the notions of 'ally', 'new object', 'actant' and 'supporter', but you have omitted the only important one, the only supporter who really counts, Nature herself. Her presence or absence explains it all. Whoever has Nature in their camp wins, no matter what the odds against them are. Remember Galileo's sentence, '1000 Demosthenes and 1000 Aristotles may be routed by any average man who brings Nature in.' All the flowers of rhetoric, all the clever contraptions set up in the laboratories you describe, all will be dismantled once we go from controversies about Nature to what Nature is. The Goliath of rhetoric with his laboratory set-up and all his attendant Philistines will be put to flight by one David alone using simple truths about Nature in his slingshot! So let us forget all about what you have been writing for a hundred pages – even if you claim to have been simply following us – and let us see Nature face to face!'

Is this not a refreshing objection? It means that Galileo was right after all. The dreadnoughts I studied in Chapters 1 and 2 may be easily defeated in spite of the many associations they knit, weave and knot. Any dissenter has got a chance. When faced with so much scientific literature and such huge laboratories, he or she has just to look at Nature in order to win. It means that there is a *supplement*, something more which is nowhere in the scientific papers and nowhere in the labs which is able to settle all matters of dispute. This objection is all the more refreshing since it is made by the scientists themselves, although it is clear that this rehabilitation of the average woman or man, of Ms or Mr Anybody, is also an indictment of these crowds of allies mustered by the same scientists.

Let us accept this pleasant objection and see how the appeal to Nature helps us to distinguish between, for instance, Schally's claim about GHRH and Guillemin's claim about GRF. They both wrote convincing papers, arraying many resources with talent. One is supported by Nature – so his claim will be made a fact – and the other is not – it ensues that his claim will be turned into an artefact by the others. According to the above objections, readers will find it easy to give the casting vote. They simply have to see who has got Nature on his side.

It is just as easy to separate the future of fuel cells from that of batteries. They both contend for a slice of the market; they both claim to be the best and most efficient. The potential buyer, the investor, the analyst are lost in the mist of a controversy, reading stacks of specialised literature. According to the above objection, their life will now be easier. Just watch to see on whose behalf Nature will talk. It is as simple as in the struggles sung in the Iliad: wait for the goddess to tip the balance in favour of one camp or the other.

A fierce controversy divides the astrophysicists who calculate the number of neutrinos coming out of the sun and Davis, the experimentalist who obtains a much smaller figure. It is easy to distinguish them and put the controversy to rest. Just let us see for ourselves in which camp the sun is really to be found. Somewhere the natural sun with its true number of neutrinos will close the mouths of dissenters and force them to accept the facts no matter how well written these papers were.

Another violent dispute divides those who believe dinosaurs to have been cold-blooded (lazy, heavy, stupid and sprawling creatures) and those who think that dinosaurs were warm-blooded (swift, light, cunning and running animals).[16] If we support the objection, there would be no need for the 'average man' to read the piles of specialised articles that make up this debate. It is enough to wait for Nature to sort them out. Nature would be like God, who in medieval times judged between two disputants by letting the innocent win.

In these four cases of controversy generating more and more technical papers and bigger and bigger laboratories or collections, Nature's voice is enough to stop the noise. Then the obvious question to ask, if I want to do justice to the objection above, is 'what does Nature say?'

Schally knows the answer pretty well. He told us in his paper, GHRH *is* this amino-acid sequence, not because he imagined it, or made it up, or confused a piece of haemoglobin for this long-sought-after hormone, but because this is what the molecule is in Nature, independently of his wishes. This is also what Guillemin says, not of Schally's sequence, which is a mere artefact, but of his substance, GRF. There is still doubt as to the exact nature of the real hypothalamic GRF compared with that of the pancreas, but on the whole it is certain that GRF is indeed the amino-acid sequence cited in Chapter 1. Now, we have got a problem. Both contenders have Nature in their camp and say what it says. Hold it! The challengers are supposed to be refereed by Nature, and not to start another dispute about what Nature's voice really said.

We are not going to be able to stop this new dispute about the referee, however, since the same confusion arises when fuel cells and batteries are opposed. 'The technical difficulties are not insurmountable,' say the fuel cell's supporters. 'It's just that an infinitesimal amount has been spent on their resolution compared to the internal combustion engine's. Fuel cells are Nature's way of storing energy; give us more money and you'll see.' Wait, wait! We were supposed to judge the technical literature by taking another outsider's point of view, not to be driven back *inside* the literature and *deeper* into laboratories.

Yet it is not possible to wait outside, because in the third example also, more and more papers are pouring in, disputing the model of the sun and modifying the number of neutrinos emitted. The real sun is alternately on the side of the theoreticians when they accuse the experimentalists of being mistaken and on the side of the latter when they accuse the former of having set up a fictional model of the sun's behaviour. This is too unfair. The real sun was asked to tell the two contenders apart, not to become yet another bone of contention.

More bones are to be found in the paleontologists' dispute where the real dinosaur has problems about giving the casting vote. No one knows for sure what it was. The ordeal might end, but is the winner really innocent or simply stronger or luckier? Is the warm-blooded dinosaur more like the real dinosaur, or is it just that its proponents are stronger than those of the cold-blooded one? We expected a final answer by using Nature's voice. What we got was a new fight over the composition, content, expression and meaning of that voice. That is, we get *more* technical literature and *larger* collections in bigger Natural History Museums, not less; *more* debates and not less.

I interrupt the exercise here. It is clear by now that applying the scientists' objection to any controversy is like pouring oil on a fire, it makes it flare anew. Nature is not outside the fighting camps. She is, much like God in not-so-ancient wars, asked to support all the enemies at once. 'Natur mit uns' is embroidered on all the banners and is not sufficient to provide one camp with the winning edge. So what is sufficient?

(2) The double-talk of the two-faced Janus

I could be accused of having been a bit disingenuous when applying scientists' objections. When they said that something more than association and numbers is needed to settle a debate, something outside all our human conflicts and interpretations, something they call 'Nature' for want of a better term, something that eventually will distinguish the winners and the losers, they did not mean to say that we know what it is. This supplement beyond the literature and laboratory trials is unknown and this is why they look for it, call themselves 'researchers', write so many papers and mobilise so many instruments.

'It is ludicrous,' I hear them arguing, 'to imagine that Nature's voice could stop Guillemin and Schally from fighting, could reveal whether fuel cells are superior to batteries or whether Watson and Crick's model is better than that of Pauling. It is absurd to imagine that Nature, like a goddess, will visibly tip the scale in favour of one camp or that the Sun God will barge into an astrophysics meeting to drive a wedge between theoreticians and experimentalists; and still more ridiculous to imagine real dinosaurs invading a Natural History Museum in order to be compared with their plaster models! What we meant, when contesting your obsession with rhetoric and mobilisation of black boxes, was that *once the controversy is settled, it is Nature the final ally that has settled it* and not any rhetorical tricks and tools or any laboratory contraptions.'

If we still wish to follow scientists and engineers in their construction of technoscience, we have got a major problem here. On the one hand scientists herald Nature as the only possible adjudicator of a dispute, on the other they recruit countless allies while waiting for Nature to declare herself. Sometimes David is able to defeat all the Philistines with only one slingshot; at other times, it is better to have swords, chariots and many more, better-drilled soldiers than the Philistines!

It is crucial for us, laypeople who want to understand technoscience, to decide which version is right, because in the first version, as Nature is enough to settle all disputes, we have nothing to do since no matter how large the resources of the scientists are, they do not matter in the end – only Nature matters. Our chapters may not be all wrong, but they become useless since they merely look at trifles and addenda and it is certainly no use going on for four other chapters to find still more trivia. In the second version, however, we have a lot of work to do since, by analysing the allies and resources that settle a controversy we understand *everything* that there is to understand in technoscience. If the first version is correct, there is nothing for us to do apart from catching the most superficial aspects of science; if the second version is maintained, there is everything to understand except perhaps the most superfluous and flashy aspects of science. Given the stakes, the reader will realise why this problem should be tackled with caution. The whole book is in jeopardy here. The problem is made all the more tricky since scientists *simultaneously* assert the two contradictory versions, displaying an ambivalence which could paralyse all our efforts to follow them.

We would indeed be paralysed, like most of our predecessors, if we were not used to this double-talk or the two-faced Janus (see introduction). The two versions are contradictory but they are not uttered by the same face of Janus. There is again a clear-cut distinction between what scientists say about the cold settled part and about the warm unsettled part of the research front. As long as controversies are rife, Nature is never used as the final arbiter since no one knows what she is and says. But *once the controversy is settled*, Nature is the ultimate referee.

This sudden inversion of what counts as referee and what counts as being refereed, although counter-intuitive at first, is as easy to grasp as the rapid passage from the 'name of action' given to a new object to when it is given its name as a thing (see above). As long as there is a debate among endocrinologists about GRF or GHRH, no one can intervene in the debates by saying, 'I know what it is, Nature told me so. It is that amino-acid sequence.' Such a claim would be greeted with derisive shouts, unless the proponent of such a sequence is able to show his figures, cite his references, and quote his sources of support, in brief, write another scientific paper and equip a new laboratory, as in the case we have studied. However, once the collective decision is taken to turn Schally's GHRH into an artefact and Guillemin's GRF into an incontrovertible fact, the reason for this decision is not imputed to Guillemin, but is immediately attributed to the independent existence of GRF in Nature. As long as the controversy lasted, no appeal to Nature could bring any extra strength to one side in the debate (it was at best an invocation, at worst a bluff). As soon as the debate is stopped, the supplement of force offered by Nature is made the explanation as to why the debate did stop (and why the bluffs, the frauds and the mistakes were at last unmasked).

So we are confronted with two almost simultaneous suppositions:

Nature is the final cause of the settlement of all controversies, *once controversies are settled.*

As long as they last *Nature will appear simply as the final consequence of the controversies.*

When you wish to attack a colleague's claim, criticise a world-view, modalise a statement you cannot *just* say that Nature is with you; 'just' will never be enough. You are bound to use other allies besides Nature. If you succeed, then Nature will be enough and all the other allies and resources will be made redundant. A political analogy may be of some help at this point. Nature, in scientists' hands, is a constitutional monarch, much like Queen Elizabeth the Second. From the throne she reads with the same tone, majesty and conviction a speech written by Conservative or Labour prime ministers depending on the election outcome. Indeed she *adds* something to the dispute, but only after the dispute has ended; as long as the election is going on she does nothing but wait.

This sudden reversal of scientists' relations to Nature and to one another is one of the most puzzling phenomena we encounter when following their trails. I believe that it is the difficulty of grasping this simple reversal that has made technoscience so hard to probe until now.

The two faces of Janus talking together make, we must admit, a startling spectacle. On the left side Nature is cause, on the right side consequence of the end of controversy. On the left side scientists are *realists*, that is they believe that representations are sorted out by what really is outside, by the only independent referee there is, Nature. On the right side, the same scientists are *relativists*, that is, they believe representations to be sorted out among themselves and the actants they represent, without independent and impartial referees lending their weight to any one of them. We know why they talk two languages at once: the left mouth speaks about settled parts of science, whereas the right mouth talks about unsettled parts. On the left side polonium was discovered long ago by the Curies; on the right side there is a long list of actions effected by an unknown actant in Paris at the Ecole de Chimie which the Curies propose to call 'polonium'. On the left side all scientists agree, and we hear only Nature's voice, plain and clear; on the right side scientists disagree and no voice can be heard over theirs.

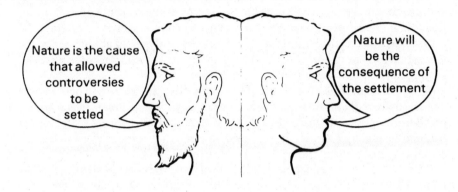

Figure 2.5

(3) The third rule of method

If we wish to continue our journey through the construction of facts, we have to adapt our method to scientists' double-talk. If not, we will always be caught on the wrong foot: unable to withstand either their first (realist) or their second (relativist) objection. We will then need to have two different discourses depending on whether we consider a settled or an unsettled part of technoscience. We too will be relativists in the latter case and realists in the former. When studying controversy – as we have so far – we cannot be *less* relativist than the very scientists and engineers we accompany; they do not *use* Nature as the external referee, and we have no reason to imagine that we are more clever than they are. For these parts of science our **third rule of method** will read: since the settlement of a controversy is *the cause* of Nature's representation not the consequence, we *can never use the outcome – Nature – to explain how and why a controversy has been settled.*

This principle is easy to apply as long as the dispute lasts, but is difficult to bear in mind once it has ended, since the other face of Janus takes over and does the talking. This is what makes the study of the past of technoscience so difficult and unrewarding. You have to hang onto the words of the right face of Janus – now barely audible – and ignore the clamours of the left side. It turned out for instance that the N-rays were slowly transformed into artefacts much like Schally's GHRH. How are we going to study this innocent expression 'it turned out'?

Using the physics of the present day there is unanimity that Blondlot was badly mistaken. It would be easy enough for historians to say that Blondlot failed because there was 'nothing really behind his N-rays' to support his claims. This way of analysing the past is called Whig history, that is, a history that crowns the winners, calling them the best and the brightest and which says the losers like Blondlot lost simply *because* they were wrong. We recognise here the left side of Janus' way of talking where Nature herself discriminates between the bad guys and the good guys. But, is it possible to use this as the reason why in Paris, in London, in the United States, people slowly turned N-rays into an artefact? Of course not, since at that time today's physics obviously could not be used as the touchstone, or more exactly since today's state is, in part, the *consequence* of settling many controversies such as the N-rays!

Whig historians had an easy life. They came after the battle and needed only one reason to explain Blondlot's demise. He was wrong all along. This reason is precisely what does not make the slightest difference while you are searching for truth in the midst of a polemic. We need, not one, but *many* reasons to explain how a dispute stopped and a black box was closed.[17]

However, when talking about a cold part of technoscience we should shift our method like the scientists themselves who, from hard-core relativists, have turned into dyed-in-the-wool realists. Nature is now taken as the cause of accurate descriptions of herself. We cannot be more relativist than scientists about these parts and keep on denying evidence where no one else does. Why? Because the cost of dispute is too high for an average citizen, even if he or she is a historian and sociologist of science. If there is no controversy among scientists as to the status of facts, then it is useless to go on talking about interpretation, representation, a biased or distorted world-view, weak and fragile pictures of the world, unfaithful spokesmen. Nature talks straight, facts are facts. Full stop. There is nothing to add and nothing to subtract.

This division between relativists and realist interpretation of science has caused analysts of science to be put off balance. Either they went on being relativists even about the settled parts of science – which made them look ludicrous; or they continued being realists even about the warm uncertain parts – and they made fools of themselves. The third rule of method stated above should help us in our study because it offers us a good balance. We do not try to undermine the solidity of the accepted parts of science. We are realists as much as the people we travel with and as much as the left side of Janus. But as soon as a controversy starts we become as relativist as our informants. However we do not follow them passively because our method allows us to document both the construction of fact and of artefact, the cold and the warm, the demodalised and the modalised statements, and, in particular, it allows us to trace with accuracy the sudden shifts from one face of Janus to the other. This method offers us, so to speak, a stereophonic rendering of fact-making instead of its monophonic predecessors!

Part II

From Weak Points to Strongholds

CHAPTER 3

Machines

Introduction: The quandary of the fact-builder

In the first part of this book we have learned how to travel through technoscience without being intimidated either by the technical literature or by the laboratories. When any controversy heats up, we know how to follow the accumulation of papers and how to take our bearings through the laboratories that stand behind the papers. To acquire this knowledge, though, we had to pay a price which can be summed up by the three principles of method I presented: first, we had to give up any discourse or opinion about science as it is made, and follow scientists in action instead; second, we had to give up any decision about the subjectivity or the objectivity of a statement based simply on the inspection of this statement, and we had to follow its tortuous history instead, as it went from hand to hand, everyone transforming it into more of a fact or more of an artefact; finally, we had to abandon the sufficiency of Nature as our main explanation for the closure of controversies, and we had instead to count the long heterogeneous list of resources and allies that scientists were gathering to make dissent impossible.

The picture of technoscience revealed by such a method is that of a weak rhetoric becoming stronger and stronger as time passes, as laboratories get equipped, articles published and new resources brought to bear on harder and harder controversies. Readers, writers and colleagues are forced either to give up, to accept propositions or to dispute them by working their way through the laboratory again. These three possible outcomes could be explored in much more detail by more studies of the scientific literature and laboratories.[1] These studies, however, no matter how necessary, would not overcome one of the main limitations of the first part of this book: dissenters are very rarely engaged in a confrontation such that, *everything else being equal*, the winner is the one with the bigger laboratory or the better article. For the sake of clarity, I started with the three outcomes above as if technoscience was similar to a boxing match. There is, in practice, a fourth set of outcomes, which is much more common: *everything*

103

not being equal, it is possible to win with many other resources than articles and laboratories. It is possible, for instance, never to encounter any dissenter, never to interest anyone, never to accept the superior strength of the others. In other words, the possession of many strongholds has first to be secured for the stronger rhetoric of science to gain any strength at all.

To picture this preliminary groundwork we have to remember our first principle: the fate of a statement depends on others' behaviour. You may have written the definitive paper proving that the earth is hollow and that the moon is made of green cheese but this paper will not become definitive if others do not take it up and use it as a matter of fact later on. You need *them* to make *your* paper a decisive one. If they laugh at you, if they are indifferent, if they shrug it off, that is the end of your paper. A statement is thus always in jeopardy, much like the ball in a game of rugby. If no player takes it up, it simply sits on the grass. To have it move again you need an action, for someone to seize and throw it; but the throw depends in turn on the hostility, speed, deftness or tactics of the others. At any point, the trajectory of the ball may be interrupted, deflected or diverted by the other team – playing here the role of the dissenters – and interrupted, deflected or diverted by the players of your own team. The total movement of the ball, of a statement, of an artefact, will depend to some extent on your action but to a much greater extent on that of a crowd over which you have little control. The construction of facts, like a game of rugby, is thus a collective process.

Each element in the chain of individuals needed to pass the black box along may act in multifarious ways: the people in question may drop it altogether, or accept it as it is, or shift the modalities that accompany it, or modify the statement, or appropriate it and put it in a completely different context. Instead of being conductors, or semi-conductors, they are all *multi-conductors*, and unpredictable ones at that. To picture the task of someone who wishes to establish a fact, you have to imagine a chain of the thousands of people necessary to turn the first statement into a black box and where each of them may or may not unpredictably transmit the statement, modify it, alter it or turn it into an artefact. How is it possible to master the future fate of a statement that is the outcome of the behaviour of all these faithless allies?

This question is all the more difficult since all the actors are doing something to the black box. Even in the best of cases they do not simply transmit it, but add elements of their own by modifying the argument, strengthening it and incorporating it into new contexts. The metaphor of the rugby game soon breaks down since the ball remains the same – apart from a few abrasions – all along, whereas in this technoscience game we are watching, the object is modified as it goes along from hand to hand. It is not only collectively transmitted from one actor to the next, it is collectively *composed* by actors. This collective action then raises two more questions. To whom can the responsibility for the game be attributed? What is the object that has been passed along?

An example will make the fact-builder's problem easier to grasp. Diesel is known as the father of the diesel engine.[2] This fatherhood, however, is not as direct as that of Athena from Zeus' head. The engine did not emerge one morning from Diesel's mind. What emerged was an idea of a perfect engine working according to Carnot's thermodynamic principles. This was an engine where ignition could occur without an increase in temperature, a paradox that Diesel solved by inventing new ways of injecting and burning fuel. At this point in the story, we have a book he published and a patent he took out; thus, we have a paper world similar to those we studied earlier. A few reviewers, including Lord Kelvin, were convinced while others found the idea impracticable.

Diesel is now faced with a problem. He needs others to transform the two-dimensional project and patent into the form of a three-dimensional working prototype. He ferrets out a few firms that build machines – Maschinenfabrik Augsburg-Nürnberg, known as MAN, and Krupp – which are interested because of the hope of increased efficiency and versatility of a perfect Carnot machine, the efficiency of the steam engine in the 1890s being pitifully low. As we will see, reality has many hues, like objectivity, and entirely depends on the number of elements tied to the claim. For four years, Diesel tried to get *one* engine working, building it with the help of a few engineers and machine tools from MAN. The progressive *realisation* of the engine was made by importing all available resources into the workshop, just as in any laboratory. The skills and tools for making pistons and valves were the result of thirty years of practice at MAN and were all locally available as a matter of routine. The question of fuel combustion soon turned out to be more problematic, since air and fuel have to be mixed in a fraction of a second. A solution entailing compressed air injection was found, but this required huge pumps and new cylinders for the air; the engine became large and expensive, unable to compete in the market of small versatile engines. By modifying the whole design of the engine many times, Diesel drifted away from the original patent and from the principles presented in his book.

The number of elements now tied to Diesel's engine is increasing. First, we had Carnot's thermodynamics plus a book plus a patent plus Lord Kelvin's encouraging comments. We now have in addition MAN plus Krupp plus a few prototypes plus two engineers helping Diesel plus local know-how plus a few interested firms plus a new air injection system, and so on. The second series is much larger, but the perfect engine of the first has been transformed in the process; in particular, constant temperature has been abandoned. It is now a constant pressure engine and in a new edition of his book Diesel has to struggle to reconcile the drift from the first more 'theoretical' engine to the one being slowly realised.

But how real is real? In June 1897 the engine is solemnly presented to the public. The worries of a black box builder now take on a new dimension. Diesel needs others to take up his engine and to turn it into a black box that runs smoothly in thousands of copies all over the world, incorporated as an unproblematic element in factories, ships and lorries. But what are these others going to do with it? How much should the prototype be *transformed* before being *transferred* from Augsburg to Newcastle, Paris or Chicago? At first, Diesel thinks that it does not have to be transformed at all: it works. Just buy the licence, pay the royalty, and we send you blueprints, a few engineers to help you, a few mechanics to tend the engine, and if you are not satisfied you get your money back! In Diesel's hands the engine is a closed black box in exactly the same way that GRF was a definitively established fact for Schally, simply waiting to be borrowed by later scientific articles (see Chapter 1).

However, this was not the opinion of the firm that had bought the prototypes. They wished it to be unproblematic, but the engine kept faltering, stalling, breaking apart. Instead of remaining closed, the black box fell open, and had to be overhauled every day by puzzled mechanics and engineers arguing with one another, exactly like Schally's readers every time they tried to get his GRF to increase the length of tibias in their own laboratories. One after the other, the licensees returned the prototypes to Diesel and asked for their money back. Diesel went bankrupt and had a nervous breakdown. In 1899, the number of elements tied to the Diesel engine *decreased* instead of increased. The reality of the engine receded instead of progressed. The engine, much like Schally's GRF, became *less* real. From a factual artefact it became, if I may use the two meanings at once, an artefactual artefact, one of those dreams the history of technics is so full of.

A few engineers from MAN, however, continued working on a new prototype. Diesel is no longer in command of their actions. A great number of modifications are made to one exemplar which operates during the day in a match factory and is overhauled every night. Each engineer adds something to the design and pushes it further. The engine is not yet a black box, but it can be made to move through more copies to many more places, undergoing incremental modifications. It is transferred from place to place without having to be redesigned. Around 1908, when Diesel's patent falls into the public domain, MAN is able to offer a diesel engine for sale, which can be bought as an unproblematic, albeit new, item of equipment and incorporated as one piece of industry. Meanwhile, the licensees who had earlier withdrawn from the project take it up, adding their contribution by designing purpose-built engines.

Just before Diesel committed suicide by jumping from a ship on the way to England, diesel engines had at last spread; but were they *Diesel's* engine? So many people had modified it since the 1887 patent that now a polemic developed about who was responsible for the collective action that made the engine real. At a 1912 meeting of the German Society of Naval Architects, Diesel claimed that it was his original engine which had been simply developed by others. However, several of Diesel's colleagues argued at the same meeting that the new real engine and the earlier patent had, at best, a weak relation, and that most of the credit should go to the hundreds of engineers who had been able to transform an unworkable idea into a marketable product. Diesel, they argued, might be the *eponym* for the collective action, but he was not the cause of this action; he was at best the inspiration, not, so to speak, the motor behind his engine.

How are we to follow these moving objects that are transformed from hand to hand and which are made up by so many different actors, before ending up as a black box safely concealed beneath the bonnet of a car, activated at the turn of a key by a driver who does not have to know anything about Carnot's thermodynamics. MAN's know-how or Diesel's suicide?

A series of terms are traditionally used to tell these stories. First, one may consider that all diesel engines lie along one *trajectory* going through different phases from ideas to market. These admittedly fuzzy phases are then given different names. Diesel's idea of a perfect engine in his mind is called *invention*. But since, as we saw, the idea needs to be developed into a workable prototype, this new phase is called *development* – hence the expression Research and Development that we will see in Chapter 4. *Innovation* is often the word used for the next phase, through which a few prototypes are prepared so as to be copied in thousands of exemplars sold throughout the world.

However, these terms are of no great use. Right from the start, Diesel had an overall notion not only of his engine, but also of the economic world in which it should work, of the way to sell licenses, of the organisation of the research, of the companies to be set up to build it. In another book Diesel even designed the type of society, based on solidarity, that would be best fit for the sort of technical novelties he wished to introduce. So no clear-cut distinction may be made between invention and innovation. In 1897 the MAN manager, Diesel and the first investors all thought that development had ended and that innovation was starting, even though it took ten more years to reach such a stage, and in the meantime Diesel went bankrupt. Thus this distinction between phases is not immediately given. On the contrary, making separations between the phases and enforcing them is one of the inventor's problems: is the black box really black? When is the dissenting going to stop? Can I now find believers and buyers? Finally, it is not even sure that the first invention should be sought in Diesel's own mind. Hundreds of engineers were looking for a more efficient combustion engine at the same time. The first flash of intuition might not be in one mind, but in many minds.

If the notion of discrete phases is useless, so, too, is that of trajectory. It does not describe anything since it is again one of the problems to be solved. Diesel indeed claimed that there was one trajectory which links his seminal patent to real engines. This is the only way for his patents to be 'seminal'. But this was disputed by hundreds of engineers claiming that the engine's ancestry was different. Anyway, if Diesel was so sure of his offspring, then why not call it a Carnot engine since it is from Carnot that he took the original idea? But since the original patent never worked, why not call it a MAN engine, or, a constant pressure air injection engine? We see that talking of phases in a trajectory is like taking slices from a pâté made from hundreds of morsels of meat. Although it might be palatable, it has no relation whatsoever to the natural joints of the animal. To use another metaphor, employing these terms would be like watching a rugby game on TV where only a phosphorescent ball was shown. All the running, the cunning, the excited players would be replaced by a meaningless zigzagging spot.

No matter how clumsy these traditional terms are in describing the building of facts, they are useful in accounting, that is for measuring how much money and how many people are invested (as we will see in the next chapter). From invention to development and from there to innovation and sale, the *money* to be invested increases exponentially, as does the *time* to be spent on each phase and the *number* of people participating in the construction. The spread in space and time of black boxes is paid for by a fantastic increase in the number of elements to be tied together. Bragg, Diesel or West (see Introduction) may have quick and cheap ideas that keep a few collaborators busy for a few months. But to build an engine or a computer for sale, you need more people, more time, more money. The object of this chapter is to follow this dramatic increase in numbers.

This increase in numbers is necessarily linked to the problem of the fact-builder: how to spread out in time and space. If Schally is the only person who believes in GRF, then GRF remains in one place in New Orleans, under the guise of a lot of words in an old reprint. If Diesel is the only person who believes in his perfect engine, the engine sits in an office drawer in Augsburg. In order to spread in space and to become long-lasting they all need (we all need) the actions of others. But what will these actions be? Many things, most of them unpredictable, which will transform the transported object or statement. So we are now in a quandary: either the others will not take up the statement or they will. If they don't, the statement will be limited to a point in time and space, myself, my dreams, my fantasies . . .. But if they do take it up, they might transform it beyond recognition.

To get out of this quandary we need to do two things at once:

　　to enrol others so that they participate in the construction of the fact;

　　to control their behaviour in order to make their actions predictable.

At first sight, this solution seems so contradictory as to look unfeasible. If others are enrolled they will transform the claims beyond recognition. Thus the very action of involving them is likely to make control more difficult. The solution to this contradiction is the central notion of **translation**. I will call translation the interpretation given by the fact-builders of their interests and that of the people they enrol. Let us look at these strategies in more detail.

Part A
Translating interests

(1) Translation one: I want what you want

We need others to help us transform a claim into a matter of fact. The first and easiest way to find people who will immediately believe the statement, invest in the project, or buy the prototype is to tailor the object in such a way that it caters to these people's **explicit interests**. As the name 'inter-esse' indicates, 'interests' are what lie *in between* actors and their goals, thus creating a tension that will make actors select only what, in their own eyes, helps them reach these goals amongst many possibilities. In the preceding chapters, for instance, we saw many contenders engaged in polemics. In order to resist their opponents' challenges they needed to fasten their position to less controvertible arguments, to simpler black boxes, to less disputable fields, gathering around themselves huge and efficient laboratories. If you were able to provide a contender with one of these black boxes, it is likely it will be eagerly seized and more rapidly transformed into a fact. Suppose, for instance, that while Diesel tinkers with his prototype, someone comes along with a new instrument that depicts on a simple indicator card how pressure changes with changing volume as the piston moves inside the cylinder so that the area on the diagram measures the work done. Diesel will jump at it, because it offers a neater way of 'seeing' how the invisible piston moves and because it graphically depicts, for everyone to see, that his engine covers a larger area than any other. The point is that, by borrowing the indicator card in order to further his goals, Diesel lends his force to its inventor, fulfilling the latter's goals. The more such elements Diesel is able to link himself to, the more likely he is to transform his own prototype into a working engine. But this movement does the same for the indicator card, which now becomes a routine part of the testing bench. The two interests are moving in the same direction.

Suppose, to take another example, that Boas, the American anthropologist, is engaged in a fierce controversy against eugenicists, who have so convinced the United States Congress of biological determinism that it has cut off the immigration of those with 'defective' genes.[3] Suppose, now, that a young anthropologist demonstrates that, at least in one Samoan island, biology cannot be the cause of crisis in adolescent girls because cultural determinism is too strong. Is not Boas going to be 'interested' in Mead's report – all the more so since he sent her there? Every time eugenicists criticise his cultural determinism, Boas will fasten his threatened position to Mead's counter-example. But every time Boas and other anthropologists do so, they turn Mead's story into more of a fact. You may imagine Mead's report interesting nobody, being picked up by no one, and remaining for ever in the (Pacific) limbo. By linking her thesis to Boas's struggle, Mead forces all the other cultural determinists to become her fellow builders: they all willingly turn her claims into one of the hardest facts of anthropology for many decades. When Freeman, another anthropologist, wished to undermine Mead's fact, he also had to link his struggle to a wider one, that of the sociobiologists. Until then, every time the sociobiologists fought against cultural determinism, they stumbled against this fact of Mead's, which had been made formidable by the collective action of successive generations of anthropologists. Sociobiologists eagerly jumped at Freeman's thesis since it allowed them to get rid of this irritating counter-example, and lent him their formidable forces (their publishing firms, their links with the media). With their help what could have been a 'ludicrous attack' became 'a courageous revolution' that threatened to destroy Mead's reputation.

As I stress in Chapter 2, none of these borrowings will be enough alone to stop the controversy: people may contest the indicator card borrowed by Diesel, or Mead's report, or Freeman's 'courageous revolution'. The point here is that the easiest means to enrol people in the construction of facts is to let oneself be enrolled by them! By pushing their explicit interests, you will also further yours. The advantage of this piggy-back strategy is that you need no other force to transform a claim into a fact; a weak contender can thus profit from a vastly stronger one.

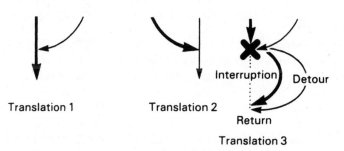

Figure 3.1

There are disadvantages as well. First, since so many people are helping you to build your claim, how will your own contribution be evaluated? Will it not be made marginal? Or worse, will it not be appropriated by others who say they did most of the work, as happened with Diesel? Second, since the contenders are the ones who have to go out of their way to follow the direction of the others (see Figure 3.1, Translation 1) they have no control on what the crowd they follow is going to do with their claims. This is especially difficult when others are so easily convinced that they turn your tentative statements into claims of gigantic size. When Pasteur elaborated a vaccine against fowl cholera that cured a few hens, he interested so many powerful groups of health officers, veterinary surgeons and farm interests that they jumped to the conclusion that 'this was the beginning of the end of all infectious diseases in men and animals'.[4] This new claim was a *composition* made in small measure from Pasteur's study of a few hens and in much larger measure from the interests of the enrolled groups. The proof that this extension was not due to Pasteur's study but rather to separate interests is that many other professions that Pasteur had not yet succeeded in interesting – the average physician for instance – found the very same experiments to be deficient, doubtful, premature and inconclusive.

Riding piggy-back is thus precarious: sometimes you have to overcome the indifference of the other groups (they refuse to believe you and to lend you their forces), and sometimes you have to restrain their sudden enthusiasm. For instance, one of the people who was not convinced by Pasteur was Koch, his German rival. But later in his career Koch had to give a lecture at the 1890

International Medical Association meeting in Berlin[5]. He had been so successful in interesting everyone in his study of tuberculosis, so clever in linking his science to the nationalism of Kaiser William, that everyone was ready to believe him. So ready indeed that when during his speech he alluded to a possible vaccine against tuberculosis everyone heard him saying that he *had* his vaccine. Everyone jumped to their feet and applauded frantically and Koch, puzzled by this collective transformation of his claim into a fact, did not dare say that he had not got a vaccine at all. When patients with tuberculosis flocked to Berlin for injections, they were bitterly disappointed, because Koch could not deliver on his ostensible promise . . . Catering to other explicit interests is not a safe strategy. There must be better ways.

(2) Translation two: I want it, why don't you?

It would be much better if the people mobilised to construct our claims were to follow us rather than the other way around. A good idea indeed, but there seems to be no reason on earth why people should go out of their way and follow yours instead (Figure 3.1, Translation 2) especially if you are small and powerless while they are strong and powerful. In fact, there is only one reason: it is *if their usual way is cut off.*

For instance, a rich businessman with an interest in philosophy wishes to establish a Foundation to study the origins of logical abilities in man. His pet project is to have scientists discover the specific neurons for induction and deduction. Talking to scientists he soon realises that they consider his dream as premature, they cannot help him reach his goal yet: but they nevertheless ask him to invest his money – now without a goal – into *their* research. He then opens a private Foundation where people study neurons, children's behaviour, rats in mazes, monkeys in tropical forests and so on . . . Scientists do what they want with his money, and not what he wanted.

This strategy, as you may see from Figure 3.1, is symmetrical with the former. The millionaire, shifting his interests, takes up those of the scientists. Such a displacement of explicit interest is not very feasible and is rare. Something else is needed to make it practical.

(3) Translation three: if you just make a short detour . . .

Since the second strategy is only rarely possible, a much more powerful one needs to be devised, as irresistible as the advice of the serpent to Eve: 'You cannot reach your goal straight away, but if you come my way, you would reach it *faster*, it would be a short cut.' In this new rendering of others' interests, the contenders do not try to shift them away from their goals. They simply offer to guide them through a short cut. This is appealing if three conditions are fulfilled: the main road is clearly cut off; the new detour is well signposted; the detour appears short.

The brain scientists would never have answered in the way I suggested when probed by the businessman above. On the contrary, they would have argued that the millionaire's goal is indeed attainable, but not right now. A little detour through *their* neurology is necessary for a few years before the neurons of induction and deduction which he is aiming at are eventually discovered. If he agrees to finance studies on acetylcholine behaviour in two synapses, he will soon be able to understand human logical abilities. Just follow the guide and be confident.

At the beginning of this century naval architects had learned to build bigger and stronger battleships by using more and more steel. However, the magnetic compasses of these dreadnoughts went wild with so much iron around. Even though they were stronger and bigger, the battleships were on the whole *weaker* than before since they got lost at sea.[6] It was at this point that a group with a solution, led by Sperry, suggested that naval architects give up the magnetic compass and use instead gyrocompasses that did not depend on magnetic fields. Did they have the gyrocompass? Not quite. It was not yet a black box offered for sale: this is why a detour had to be negotiated. The Navy must invest in Sperry's research in order to convert his idea into a workable gyroscope, so that, in the end, their battleships can steer a straight course again. Sperry has positioned himself so that a common translation of his interests and that of the Navy now reads: 'You cannot navigate your ships properly; I can't make my gyrocompass a real thing; wait a little, come my way, and after a while your ships will make full use of their terrifying powers again and my gyrocompasses will spread in ships and planes in the form of well-closed black boxes.'

This community of interests is the result of a difficult and tense negotiation that may break down at any point. In particular, it is based on a sort of implicit contract: there should be a return to the main road, and the detour should be short. What happens if it becomes long, so long indeed that it now appears in the eyes of the enrolled groups as a deviation rather than a short cut? Imagine that for a decade the millionaire keeps reading papers on the firing of synapses, expecting the discovery of the neurons for induction and deduction any day. He might die of boredom before seeing his dreams fulfilled. He might think that this is not the detour they had agreed upon, but a new direction altogether. He might even realise that it is the *second strategy* which has been practised, not the third and then decide to sever the negotiations, to cut the money off, and to dismiss the scientists who were not only pulling his leg but also using his money.

This is what occurred with Diesel. MAN was ready to wait for a few years, to lend engineers, with the idea that they would soon resume their usual business of manufacturing engines but on a larger scale. If the return is delayed, the management may feel cheated, as if they were perceiving the second type of translation through the veil of the third. If they start thinking this way, then Diesel is taken as a parasite on MAN diverting its resources to further his own egotistical dreams. Interests are elastic, but like rubber, there is a point where they break or spring back.

So, even if this third way of translating the interest of others is better than the second, it has its shortcomings. It is always open to the accusation of bootlegging – to use the expression of American scientists – that is, the size of the detour and the length of the delay being fuzzy, a detour might be seen as an outright diversion, or even as a hijacking. Support may thus be cut off *before* Watson and Crick discover the double helix structure, Diesel has time to make his engine, West to build his *Eagle* computer, Sperry his gyrocompass, and the brain scientists to find how a synapse fires. There is no accepted standard for measuring detours because the 'acceptable' length of the detour is a result of negotiation. MAN, for instance, became worried after only a few years, but the private medical foundations that invested in Lawrence's huge accelerators at Berkeley did not, even though Lawrence was furthering particle physics by arguing that he was building bigger radiation sources for cancer therapy![7] Depending on the negotiators' abilities, a few hundred dollars may appear to be an intolerable waste of money, while building cyclotrons looks like the only straight path to a cure for cancer.

There are two other limitations to this third strategy. First, whenever the usual road is not blocked, whenever it is not clearly apparent to the eyes of a group that they cannot follow their usual route, it becomes impossible to convince them to make a detour. Second, once the detour has been completed and everyone is happy, it is very hard to decide who is responsible for the move. Since the Navy helped Sperry, it can claim credit for the whole gyrocompass, which would otherwise have remained a vague sketch or an engineer's blueprint. But since without his gyrocompass the Navy feared that its dreadnoughts would be lost at sea, Sperry may very well claim to be the active force behind the Navy. There may be a bitter struggle to allocate credit, even when everything goes well.

(4) Translation four: reshuffling interests and goals.

A fourth strategy is needed to overcome the shortcomings of the third:
(a) the length of the detour should be impossible to evaluate for those who are enlisted;
(b) it should be possible to enrol others even if their usual course is not obviously cut off;
(c) it should be impossible to decide who is enlisted and who does the enlisting;
(d) nevertheless, the fact-builders should appear as the only driving force.

To carry off what would seem to be a quite impossible task, there is one obstacle that seems at first to be unsurmountable: people's *explicit* interests. So far I have used the term 'explicit interest' in a non-controversial way: the Navy has interests, so has the millionaire, so has MAN, so have all the other actors we have followed. All of them know more or less what they want, and a list of their goals may, at least in principle, be set up, either by them or by observers. As long as the goals of all these actors are explicit, the fact-builder's degree of freedom is limited to the narrow circle delineated by the three strategies above. The enlisted groups know that they are a group; know where they want to go; know if their usual way is interrupted; know how far they are ready to deviate from it; know when they have returned to it; and finally, know how much credit should go to those who helped them for a while. They know a lot![8] They know too much because this knowledge limits the moves of the contenders and paralyses negotiations. As long as a group possesses such knowledge, it will be extremely hard to enrol it in the fact-building and, still more, to control its behaviour. But how to bypass this obstacle? The answer is simple and radical. By following fact-builders in action we are going to see one of their most extraordinary feats: they are going to *do away* with explicit interests so as to increase their margin for manoeuvre.

(A) TACTIC ONE: DISPLACING GOALS

Even if they are explicit, the meanings of people's goals may be differently interpreted. A group with a solution is looking for a problem but no one has a problem....Well, why not make them have a problem? If a group feels that its usual way is not at all interrupted, is it not possible to offer it another scenario in which it has got a big problem?

When Leo Szilard first entered into discussion with the Pentagon in the early 1940's, the generals were not interested in his proposal to build an atomic weapon[9]. They argued that it always takes a generation to invent a new weapon system, that putting money into this project might be good for physicists for doing physics but not soldiers for waging war. Thus they saw Szilard's proposal as a typical case of bootlegging: physicists would be better occupied perfecting older weapon systems. Since they did not feel their usual way of inventing weapons was cut off, the generals had no reason to see Szilard's proposal as a solution to a non-existent problem. Then Szilard started to work on the officers' goals. 'What if the Germans got the atom bomb first? How will you manage to win the war – your explicit aim – with all your older and obsolete weapons?' The generals had to win a war–'a war' in its usual rendering means a classical one: after Szilard's intervention they still had to win the war – meaning now a new atomic one. The shift in meaning is slight but sufficient to change the standing of the atomic physicists: useless in the first version, they become *necessary* in the second. The war machine is not being invaded by bootlegging physicists any more. It is now geared full speed towards the progressive realisation of Szilard's vague patent into a not so vague bomb...

(B) TACTIC TWO: INVENTING NEW GOALS

Displacing the goals of the groups to be enlisted so as to create the problem and then offering a possible solution is nice, but still limited by the original aims. Thus, in this example, Szilard could convince the Pentagon to wage a nuclear war, but not to lose it or to support classical dance. The margin of freedom would be much increased if new goals could be devised.

When George Eastman tried to move into the business of selling photographic plates, he soon realised that he could convince only a few, well-equipped amateurs to buy his plates and his paper[10]. They were used to working in semi-professional laboratories built in their homes. Others were *uninterested* in taking pictures themselves. They did not want to buy costly and cumbersome black boxes – this time in the literal sense of the word! Eastman then devised the notion of 'amateur photography': everyone from 6 to 96 years old might, could, should, want to take photographs. Having this idea of a mass market, Eastman and his friends had to define the object that would convince everybody to take photographs. Only a few people were ready for a long detour through expensive laboratories. The Eastman Company had to make the detour extremely small to enlist everyone. So that no one should hesitate to take a picture, the object should be cheap, and easy, so easy that, as Eastman put it: 'You press the button, we do the rest', or as we say in French, 'Clic, clac, merci Kodak'. The camera was not yet there, but Eastman already sensed the contours of the object which would make his company indispensable. Previously few people had had the goal of taking photographs. If Eastman was successful, everyone would have this goal, and the only way to fulfil this craving would be to buy camera and films from the local Eastman Company dealer.

(C) TACTIC THREE: INVENTING NEW GROUPS

This is easier said than done. Interests are the consequence of whatever groups have been previously engaged to do. MAN builds steam engines; it may be persuaded to build diesel engines, but not easily persuaded to make yoghurt. The Pentagon wishes to win the war; they might be persuaded to win an atomic one, but not easily to dance, and so on. The ability to invent new goals is *limited* by the existence of already defined groups. It would be much better to *define* new groups that could then be *endowed* with new goals, goals which could, in turn, be reached only by helping the contenders to build their facts. At first sight, it seems impossible to invent new groups; in practice, it is the easiest and by far the most efficient strategy. For instance, Eastman could not impose a new goal – taking pictures – without devising a new group from scratch, the amateur photographer from age 6 to 96.

In the mid-nineteenth century, rich and poor, capitalist and proletariat were some of the most solidly defined groups because of the class struggle. Health officers who wished to overhaul European and American cities to make them safe and hygienic were constantly stalled by class hostility between poor and rich[11]. The simplest regulation for health was considered either to be too radical, or, on the contrary, to be one more stick for the rich to beat the poor with. When

Pasteur and the hygienists introduced the notion of a microbe as the essential cause of infectious disease, they did not take the society to be made up of rich and poor, but of a rather different list of groups: sick contagious people, healthy but dangerous carriers of the microbes, immunised people, vaccinated people, and so on. Indeed, they added a lot of *non-human actors* to the definition of the groups as well: mosquitoes, parasites, rats, fleas, plus the millions of ferments, bacteria, micrococci and other little bugs. After this reshuffling, the relevant groups were not the same: a very rich man's son could die simply because the very poor maid was carrying typhoid. As a consequence, a different type of solidarity emerged. As long as society was made up just of classes, hygienists did not know how to become indispensable. Their advice was not followed, their solutions were not applied. As soon as newly formed groups were threatened by the newly invented enemy, common interest was created, and so was a craving for the biologists' solutions; hygienists allied with microbiologists were positioned at the centre of all regulations. Vaccines, filters, antiseptics, know-how that had until then been confined to a few laboratories spread to every household.

(D) TACTIC FOUR: RENDERING THE DETOUR INVISIBLE

The third tactic has its shortcomings as well. As long as a group – even made up – is able to detect a widening gap between its goals – even displaced – and that of the enrolling groups the margin of negotiation of the latter is much restrained. People can still *see* the difference between what they wanted and what they got, they still can feel they have been cheated. A fourth move is thus necessary that turns the detour into a progressive drift, so that the enrolled group still thinks that it is going along a *straight* line without ever abandoning its own interests.

In Chapter 1 we studied such a drift. The managers of a big company were after new, more efficient, cars. They had been convinced by their research group that electric cars using fuel cells were the key to the future. This was the first translation: 'more efficient cars' equals 'fuel cells'. But since nothing was known of fuel cells they were convinced by the research director that the crucial enigma to be tackled was the behaviour of electrodes in catalysis[12]. This provided the second translation. The problem, they were later told by engineers, was that the electrode is so complex that they should study a single pore of a single electrode. The third translation now reads: 'study of catalysis' = 'study of one pore' (see Chapter 1, sentence (8)). But since the series of translations is a *transitive* relation the final version upheld by the Board of Directors was: 'new efficient cars' = 'research into the one-pore model'. No matter how far the drift might appear, it is not felt as a detour any more. On the contrary, it has become the only *straight* way to get at the car. The Board's interests have to go through this one pore like the camel through the eye of the needle!

To take another example, a French columnist argued, in 1871, after the Franco-Prussian war, that if the French had been beaten, this was due to the German soldiers' better state of health. This is the first translation that offers a

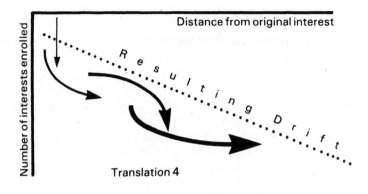

Figure 3.2

new rendering of the military disaster. Then he goes on by arguing that this better health was due to German superiority *in science*. Translation two expounds a new interpretation of the usefulness of basic science. He then explained that science was superior in Germany because it was better funded. Third translation. He next tells the reader that the French Assembly was, at that moment, cutting funds for basic science. This makes for a fourth displacement: no revenge would ever be possible if we had no money, since there is no science without money, no healthy soldiers without science and no revenge without soldiers. In the end he suggests to the reader what to do: write to your representative to make him change his vote. All the slight displacements are smoothly nested, one in another, so that the same reader who was ready to pick up his rifle and march on the Alsatian frontier to beat the Germans, was now, *with the same energy*, and *without* having eschewed his goal, writing an indignant letter to his representative!

It should now be clear why I used the word *translation*. In addition to its linguistic meaning (relating versions in one language to versions in another one) it has also a geometric meaning (moving from one place to another). Translating interests means at once offering new interpretations of these interests and channelling people in different directions. 'Take your revenge' is made to mean 'write a letter'; 'build a new car' is made to really mean 'study one pore of an electrode'. The results of such renderings are a slow movement from one place to another. The main advantage of such a slow mobilisation is that particular issues (like that of the science budget or of the one-pore model) are now *solidly tied* to much larger ones (the survival of the country, the future of cars), so well tied indeed that threatening the former is tantamount to threatening the latter. Subtly woven and carefully thrown, this very fine net can be very useful at keeping groups in its meshes.

(E) TACTIC FIVE: WINNING TRIALS OF ATTRIBUTION

All the above moves enormously increase the contender's room for manoeuvre, especially the latter, which dissolves the notion of explicit interest. It is no longer possible to tell who is enrolled and who is enrolling, who is going out of his way and who is not. But this success brings its own problems with it. How can we decide who did the job, or indeed, how can the fact-builders determine if the facts eventually built are *their own?* All along we meet this problem: with Diesel's engine, with Pasteur's vaccine, with Sperry's gyrocompass. The whole process of enrolment, no matter how cleverly managed, may be wasted if others gain credit for it. Conversely, enormous gains may be made simply by solving it, even if the process of enrolment has been badly managed.

After reading a famous work by Pasteur on fermentation, an English surgeon, Lister, 'had the idea' that wound infections – that killed most if not all of his patients – might be similar to fermentation[13]. Imitating Pasteur's handling of fermenting wine, Lister then imagined that by killing the germs in the wounds and by letting oxygen pass through the dressing, infection would stop and the wound heal cleanly. After many years of trials, he invented asepsis and antisepsis. Hold on! Did he invent them? A new discussion starts. No he did not, because many surgeons had had the idea of linking infection and fermentation before, and of letting air through the bandage; many colleagues worked with and against him for many years before asepsis became a routine black box in all surgical wards. Besides, in many lectures Lister gracefully *attributed* his original ideas to Pasteur's memoir. So, in a sense he 'simply developed' what was in germ, so to speak, in Pasteur's invention. But Pasteur never made asepsis and antisepsis a workable practice in surgery; Lister did. So, in another sense, Lister did everything. Historians, as much as the actors themselves, delight in deciding who influenced whom, who had only a marginal contribution and who made the most significant contribution. With each new witness, someone else, or some other group, takes credit for part or for all of the move.

So as not to be confused, we should distinguish the recruiting of allies so as to build a fact or a machine collectively, from the *attributions of responsibility* to those who did most of the work. By definition, and according to our first principle, since the construction of facts is collective, everyone is as necessary as anyone else. Nevertheless, it is possible, in spite of this necessity, to make everyone accept a few people, or even one person, as the main cause for their collective work. Pasteur, for instance, not only recruited many sources of support, but also strove to maintain his laboratory as the source of the general movement that was made up of many scientists, officials, engineers and firms. Although he had to accept their views and follow their moves – so as to extend his lab – he also had to fight so that they all appeared as simply 'applying' his ideas and following his leads. The two movements must be carefully distinguished because, although they are complementary for a successful strategy, they lead in opposite directions: the recruitment of allies supposes that you go as far and make as many compromises as possible, whereas the attribution of responsibility requires you to *limit* the number of actors as much as possible. The question of knowing who follows and who is followed should in no way be asked if the first movement is to succeed, and nevertheless should be settled for the second movement to be completed. Although Diesel followed many of the people he recruited, translating their common interest in an ambiguous mixture, in the end he had to make them consider his science as the leader *they followed.*

I will call the **primary mechanism** that which makes it possible to solve the enrolment problem and make the collective action of many people turn from 'germs' into reality asepsis, gyrocompasses, GRF or diesel engines. To this mechanism a **secondary mechanism** has to be added which might have no relation at all with the first and which is as controversial and as bitter as the other ones.

A military metaphor will help us remember this essential point. When an historian says that Napoleon *leads* the Great Army through Russia every reader knows that Napoleon with his own body is not strong enough to win, say, the battle of Borodino[14]. During the battle half a million people are taking initiatives, mixing up the commands, ignoring orders, fleeing or courageously dying. This gigantic mechanism is much bigger than what Napoleon can handle or even see from the top of a hill. Nevertheless, after the battle, his soldiers, the Tsar, Kutuzov who commands the Russian army, the people of Paris, the historians, all attribute to him and only to him the responsibility for the victory – which in this case turned out later to be a defeat. Everyone will agree that there may be *some* relation between what Napoleon did during the battle and what the hundreds of thousands of others did, but they will also agree that these relations cannot be captured by the sentence 'Napoleon won because he had the power and the others obeyed'. Exactly the same is true of the relations between the handful of scientists and the millions of others. Their complicated and unpredictable relations cannot be captured by a simple order of command that would go from basic science to the rest of society via applied science and development.

Other people will decide that Diesel was a mere precursor, or that Pasteur did all the basic work on asepsis, or that Sperry had only a marginal input into the gyrocompass. Even when all these questions are later tackled by historians, their research *adds* an important expert testimony to the trials, but it does not *end* the trials and does not take the place of the court. In practice, however, people make some versions more credible than others. Everyone may finally accept that Diesel 'had the idea' of his engine, that Lister 'invented' asepsis with the help of Pasteur's memoirs, or that Napoleon 'led' the Great Army. For a reason that will become still clearer in Part C, this secondary distribution of flags and medals should never be confused with the primary process.

(5) Translation five: becoming indispensable

The contenders now have a lot of leeway with these five tactics in their attempts to interest people in the outcome of their claims. With guile and patience it should be possible to see everyone contributing to the spread of a claim in time and space – which will then become a routine black box in everyone's hands. If such a point were reached, then no further strategy would be necessary: the contenders would have simply become *indispensable*. They would not have to cater to others' interests – first translation – nor to convince them that their usual ways were cut off – second translation – nor seduce them through a little detour – third translation; it would no longer even be necessary to invent new groups, new goals, to surreptitiously bring about drift in interests, or to fight bitter struggles for attribution of responsibilities. The contenders would simply sit at a particular place, and the others would flow effortlessly through them, borrowing their claims, buying their products, willingly participating in the construction and spread of black boxes. People would simply rush to buy Eastman Kodak cameras, to have Pasteur's injections, to try Diesel's new engines, to install new gyrocompasses, to believe Schally's claims without a shadow of a doubt, and to dutifully acknowledge the ownership rights of Eastman, Pasteur, Diesel, Sperry and Schally.

The quandary of the fact-builder would not simply be precariously patched up. It would be entirely resolved. No negotiation, no displacement would be necessary since *the others would do the moving*, the begging, the compromising and the negotiation. They are the ones who would go out of their way. In Figures 3.1 and 3.2 I pictured the four translations. They all lead to the fifth translation that literally sums them up. In the geometric sense of translation it means that whatever you do, and wherever you go, you *have* to pass through the contenders' position and to help them further their interests. In the linguistic sense of the

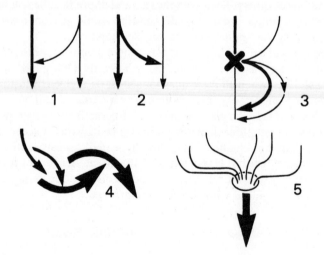

Figure 3.3

word translation, it means that one version translates every other, acquiring a sort of hegemony: whatever your want, you want this as well. The diagram makes clear that, from the first to the last, the contenders have shifted from the most extreme weakness – that forced them to follow the others – to the greatest strength – that forces all the others to follow them.

Is such a strategy feasible? Shadowing scientists and engineers will show us that it is common practice, but that, in order to succeed, other allies have to be brought in and most of them do not look like men or women.

Part B
Keeping the interested groups in line

We saw in the introduction to this chapter that two things are needed in order to build a black box: first it is necessary to *enrol* others so that they believe it, buy it and disseminate it across time and space; second, it is necessary to *control* them so that what they borrow and spread remains more or less the same. If people are not interested, or if they do something entirely different with the claim, the spread of a fact or of a machine in time and space does not take place. A few people toy with an idea for a few days, but it soon disappears, to be replaced by another. Projects which trigger enthusiasm are quickly put back into a drawer. Theories that had started to infect the world shrink back to become the *idée fixe* of some lunatic in an asylum. Even colleagues who had been 'unalterably' convinced by a laboratory demonstration can change their minds a month later. Established facts are quickly turned into artefacts, and puzzled people ask, 'How could we have believed such an absurdity?' Established industries that looked as if they were to last for ever suddenly become obsolete and start falling apart, displaced by newer ones. Dissenters who interrupt the spread of any fact or artefact proliferate.

In Part A we have seen how to do half the job, that is, how to interest others. Now we have to tackle the other half: how to make their behaviour predictable. This is a much harder task.

(1) A chain is only as strong as its weakest link

Let us first assess the difficulty of the task. When Diesel succeeded in interesting MAN in his project for a perfect engine, he was lent money, workshops, assistants, and was granted some time. His problem was to hold those elements together with the ones he was bringing into the contract: Carnot's thermodynamics, the principle of ignition at constant temperature and his own views on the future market. Initially all these elements are simply *assembled* in one place at Augsburg. What could bind them more firmly together? A working prototype which might later be used as *a single piece* of standard equipment in other settings – a submarine or a truck, for example. What will happen if Diesel cannot hold all these elements at once? The answer is simple: they will be *disbanded* as easily as they have been assembled. Each of the elements will go its own way: MAN will go on building steam engines, assistants will be moved to other jobs, money will flow elsewhere, Carnot's thermodynamics will remain a cryptic piece of basic physics, ignition at constant temperature will be remembered as a technological dead end, and Diesel will occupy himself with other tasks, leaving little trace in the history books.

So the number of enlisted interests is important but far from enough, because knitting and tying them together may be undone. Pasteur had been able to convince farmers who raised cattle that the only way to solve the terrible anthrax plague was to pass through his laboratories at the Ecole Normale Supérieure in Rue d'Ulm in Paris. Breathing down Pasteur's neck were thousands of interests nested into one another, all ready to accept his short cut through the microscope, the artificial culture of microbes, and the promised vaccine. However there is a considerable drift between an interest in raising cattle on a farm and watching microbes grow in Petri dishes: the gathering crowd might disband rapidly. After a few months of hope they might all leave disappointed, bitterly accusing Pasteur of having fooled them by creating artefacts in his laboratory of little relevance to farms and cattle. Pasteur would then become a mere precursor for the anthrax vaccine, his role in history being accordingly diminished. Something else is needed to tie the diverted resources and *invested* interests together in a durable way.

Eastman had the bright idea of inventing a new group of 6- to 96-year-olds that was endowed with a craving for taking pictures. This enlistment depended on a camera that was simple to operate, which meant a camera with film and not the expensive fragile and cumbersome glass plates then used. But what would happen if the film slackened so much that all the pictures were fuzzy? What if the coating of the film blistered? No matter how many people found photography appealing, no matter how big the Eastman Company, not matter how clever and interested Eastman might be, the associated interests would disassociate. Eastman, with his dream of a mass market, would become one of the many precursors in the long history of popular photography. Others would take up his patents, even perhaps buy his company.

Something more is needed to turn the temporary juxtaposition of interests into a durable whole. Without this 'little something', the assembly of people necessary to turn a claim into a black box will behave unpredictably: they will dissent, they will open it, tinker with it; worse, they will lose interest and drop it altogether. This 'dangerous' behaviour should be made impossible; even better, it should be made unthinkable.

We know the answer since we have been talking about it for three chapters: the only way to keep the dissenters at bay is to link the fate of the claim with so many assembled elements that it resists all trials to break it apart.

The first prototype that Diesel assembles is much like Schally's GRF or

Blondlot's ill fated N-Rays: each new trial makes it falter. At the start, Diesel ties the fate of his engine to that of *any* fuel, thinking that they would all ignite at a very high pressure. This, to him, is what made his engine so versatile. He needs very high pressure to obtain such a result, with pistons, cylinders and valves strong enough to withstand more than 33 atmospheres. MAN was able to provide him with excellent machine tools and know-how so that it soon became possible to obtain such high pressure. But then, nothing happened. Not every fuel ignited. This ally which he had expected to be unproblematic and faithful betrayed him. Only kerosene ignited, and then only erratically. How could the ignition of kerosene be kept in line? Diesel discovered that it depended on the right mixture of air and fuel. To keep this mixture constant he had to introduce the fuel and the air into the cylinder at a very high pressure. But Diesel had to add powerful pumps, sturdy valves and a lot of extra plumbing to his original design to obtain such a result. His engine may run, but it becomes large and expensive.

So what is happening? Diesel has to *shift his system of alliances:* high pressure plus any fuel plus solid injection lead to engines of any size which interest everyone and spread everywhere. But this series of associations is dismantled in the Augsburg workshop, as soon as it is tried out. The engine does not even turn one stroke. So, a new series of alliances is tried out: high pressure plus kerosene plus air injection which means a large and costly engine that idles for a few seconds.

I hear the reader's objection: 'But do we really have to go into these details to understand how others are to be controlled?' Yes, because without these little details *others are not controlled!* Like the dissenter of Chapter 2, they apply pressure to the new design, and the whole thing breaks apart. To resist dissent, that is to resist trials of force, Diesel has to invent an injection pump that holds air and kerosene together, allows the high pressure to ignite the mixture, makes the engine run, and thus keeps MAN in line. But if the kerosene, the air, and MAN are kept in line, this is not the case for the vast market anticipated by Diesel. This has to be given up. Groping in the dark inside his workshop, Diesel has to choose alliances. He has to decide what he *most* wishes to keep in line. There is at first no engine that can ally itself to air, to any fuel and to everyone's needs. *Something has to give way:* a fuel, the kerosene, solid injection, Carnot's principles, the mass market, Diesel's stamina, MAN's patience, rights to patents . . . Something.

The same choice goes on in Pasteur's laboratory. Is there anything that can be used to tie in the farmers' interests before they all go away bitter and scornful? A tiny bacillus inside a urine medium will not do, even if it is visible under the microscope. It is only of marginal interest to people who have been attracted to the lab by the promise that they will soon be back on their farms, milking healthier cows and shearing healthier sheep. If Pasteur was using his bacillus to do biochemistry or taxonomy, deciding if it was an animal or a lichen, others like biochemists or taxonomists would be interested, but not the farmers. When Pasteur shows that sheep fed older cultures of the bacillus resist the disease even when they are later fed virulent cultures, biochemists and taxonomists are only casually interested but farmers are very interested. Instead of losing interest, they gain it. This is a vaccine to prevent infection, something easy to relate to the farm conditions. But what if the vaccine works erratically? Again, interest may slacken and disappointment return. Pasteur then needs a new reliable method to turn the production of vaccine into a routine, a black box that may be injected by any vet. His collaborators discover that it all depends on the temperature of the culture: 44° for a few days is fine, the culture ages and may be used as a vaccine; at 45°, the bacillus dies; at 41° it changes form, sporulates and becomes useless as a vaccine. These little details are what clamp together the wavering interests of the enrolled farmers. Pasteur has to find ways to make *both* the farmers and the bacillus predictable. And he has to keep on discovering new ways, or at least for as long as he wishes to tie these farmers and these microbes together. The tiniest loose end in this *lash-up*[15] and all his efforts are wasted.

The captation of people's interests, and their translation to make them work in the construction of a black box, leads, I have to admit, to trifles. But if you build a long chain, it still remains only as strong as its weakest link no matter how grandiose some of its elements may be. Little matter that Eastman has mobilised his whole company to capture the amateur market; little matter that he has invented a new box, a new roller, a new film, a new ratchet for the new spring holding the negatives; if the coating of the film *blisters*, then that is the end of the whole enterprise. There is one missing link in the long chain[16]. One negligible ally defaults. Shifting from paper to celluloid allows Eastman to solve these irritating blisters. This part of the camera at least becomes indisputable. The camera now moves from hand to hand as *one object*, and may start to interest the people it has been devised to interest. Now attention shifts to another missing link, to the new machines that have to be invented to make long strips of celluloid. To keep them in line, other allies have in turn to be fetched and assembled, and so on.

(2) Tying up with new unexpected allies

We now start to understand that there is no way of tying together interested groups – mobilised in part A – unless other elements are tied with them: piston, air, kerosene, urine medium, microbes, roller, coating, celluloid, etc. But we also understand that it is not possible to tie any element to another at random. Choices have to be made. Diesel's decision to go in for air injection means that many potential buyers have to be abandoned and that Carnot's principles may not be that easily applied. Pasteur's search for a new medium for his vaccine entails the abandonment of other interests in biochemistry and taxonomy. Amateurs may be captured by Eastman's new Kodak camera, but the semi-professionals who do their own plates and development are left to one side and the new film coating had better not blister. As in Machiavelli's *Prince*, the progressive building up of an empire is a series of decisions about alliances: With whom can I collaborate? Whom should I write off? How can I make this one faithful? Is this other one reliable? Is this one a credible spokesperson? But what did not occur to Machiavelli is that these alliances can cut across the boundaries between human beings and 'things'. Every time an ally is abandoned, replacements need to be recruited; every time a sturdy link disrupts an alliance that would be useful, new elements should be brought in to break it apart and make use of the dismantled elements. These 'machiavellian' strategies are made more visible when we follow scientists and engineers. Rather, we call 'scientists' and 'engineers' those subtle enough to include in the same repertoire of ploys human and non-human resources, thus increasing their margin for negotiation.

Take for instance the Bell Company[17]. Telephone lines in the early days were able to carry a voice only a few kilometres. Beyond this limit the voice became garbled, full of static, inaudible. The message was corrupted and not transmitted. By 'boosting' the signals every thirteen kilometres, the distance could be increased. In 1910, mechanical repeaters were invented to relay the message. But these costly and unreliable repeaters could be installed only on a few lines. The Bell Company was able to expand, but not very far, and certainly not through the desert, or the Great Plains of the United States where all sorts of small companies were thriving in the midst of complete chaos. Ma Bell, as it is nicknamed by Americans, was indeed in the business of linking people together, but with the mechanical repeater many people who might wish to pass through her network could not do so. An exhibition in San Francisco in 1913 offered Bell a challenge. What if we could link the West and the East Coast with one telephone line? Can you imagine that? A transcontinental line tying the US together and rendering Bell the indispensable go-between of a hundred million people, eliminating all the small companies? Alas, this is impossible because of the cost of the old repeater. It becomes the missing link in this new alliance planned between Ma Bell and everyone in the US. The project falls apart, becomes a dream. No transcontinental line for the time being. Better send your messages through the Post Office.

Jewett, one of the directors of Bell, looks for new possible alliances that will help the company out of its predicament. He remembers that he was taught by Millikan, when the latter was a young lecturer. Now a famous physicist, Millikan works on the electron, a new object at the time, that is slowly being built up in his laboratory like all the other actants we saw in Chapter 2. One of the features of the electron is that it has little inertia. Jewett, who himself has a doctorate in physics, is ready for a little detour. Something which has no inertia loses little energy. Why not ask Millikan about a possible new repeater? Millikan's laboratory has nothing to offer, yet. Nothing ready for sale. No black box repeating long-distance messages cheaply and safely. What Millikan can do, however, is to lend Jewett a few of his best students, to whom Bell offers a well-equipped laboratory. At this point Millikan's physics is in part connected with Bell's fate, which is partly connected with the challenge of the San Francisco fair, according to a chain of translations like the ones we studied above. Through a series of slight displacements, electrons, Bell, Millikan and the continental line are closer to one another than they were before. But it is still a mere juxtaposition. The Bell Company managers may soon realise that basic physics is good for physicists but not for businessmen; electrons may refuse to jump from one electrode of the new triodes to the next when the tension gets too high, and fill the vacuum with a blue cloud; the urge for a transcontinental line may no longer be felt by the Board of Directors.

This mere juxtaposition is transformed when Arnold, one of the recruited physicists, transforms a triode patented by another inventor. In a very high vacuum, even at very high tension, the slightest vibration at one end triggers a strong vibration at the other. A new object is then created through new trials in the newly opened laboratory: electrons that greatly amplify signals. This new electronic repeater is soon transformed into a black box by the collective work of Ma Bell, and incorporated as a routine piece of equipment in six locations along the 5500 kilometres of cable laid across the continent. In 1914, the transcontinental line, impossible with the other repeater, becomes real. Alexander Bell calls Mr Watson, who is no longer downstairs but thousands of miles away. The Bell Company is now able to expand over the whole continent: consumers who had not before had the slightest interest in telephoning the other coast now routinely do so, passing through the Bell network and contributing to its expansion – as anticipated from the fifth translation described above. But the boundaries of physics have been transformed as well, from a few modestly equipped laboratories in universities to many well-endowed laboratories in industry; from now on many students could make a career in industrial physics. And Millikan? He has changed too, since many effects first stabilised in his lab are now routinely used along telephone lines, everywhere, thus providing his laboratory with a fantastic expansion. Something else has moved too. The electrons. The list of actions that defined their being has been dramatically increased when all these laboratories submitted them to new and unexpected trials. Domesticated electrons have been made to play a role in a convoluted alliance that allows the Bell Company to triumph over its rivals. In the end, each actor in this little story has been pushed out of its usual way and made to be different, because of the new alliances it has been forced to enter.

We, the laypeople, far away from the practice of science and the slow build-up of artefacts, have no idea of the versatility of the alliances scientists are ready to make. We keep nice clean boundaries that exclude 'irrelevant' elements: electrons have nothing to do with big business; microbes in laboratories have nothing to do with farms and cattle; Carnot's thermodynamics is infinitely far from submarines. And we are right. There is at first a vast distance between these elements; at the beginning they are indeed irrelevant. But 'relevance', like everything else, can be *made*. How? By the series of translations I have sketched. When Jewett first fetches Millikan, the electrons are too feeble to have any easy connection with Ma Bell. At the end, inside the triode redesigned by Arnold, they reliably transmit Alexander Bell's order to Mr Watson. The smaller companies might have thought that Ma Bell would never beat them since it was impossible to build a transcontinental line. This was counting *without* the electrons. By adding electrons and Millikan and his students and a new lab to the list of its allies, Ma Bell modifies the relations of forces. Where it was weak over longer distances, it is now stronger than anyone else.

We always feel it is important to decide *on the nature of the alliances:* are the elements human or non-human? Are they technical or scientific? Are they objective or subjective? Whereas the only question that really matters is the following: *is this new association weaker or stronger than that one.* Veterinary science had not the slightest relation with the biology done in laboratories when Pasteur began his study. This does not mean that this connection cannot be built. Through the establishment of a long list of allies, the tiny bacillus attenuated by the culture has a sudden bearing on the interests of farmers. Indeed, it is what definitively reverses the balance of power. Vets with all their science now have to pass through Pasteur's laboratory and borrow his vaccine as an incontrovertible black box. He has become indispensable. The fulfilment of the strategies presented in Part A is entirely dependent on the new unexpected allies that have been *made to be relevant.*

The consequence of these bold moves that enrol newly formed actors (microbes, electrons) in our human affairs is that there is no way to counteract them except by tackling these 'technical details'. Like the proof race described in Chapter 1, once it has started there is no way of avoiding the nitty-gritty since this is what makes the difference. Without building expensive laboratories that they could not afford in an attempt to attract physics and electrons back into their own camp, the small companies eliminated by Bell could not resist. The laboratories studied in Chapter 2 now occupy the centre of these strategies through which new actors constituting a vast reservoir of forces are mobilised. The spokespersons able to talk on behalf of new and invisible actors are now the linchpins on which the balance of power rests: a new characteristic of electrons, one more degree in the culture medium, and the whole assembled crowd either breaks up or is irreversibly bound.

The intimate details of an obscure science may become a battlefield like a hitherto modest hamlet became the stage for the battle of Waterloo. In Edinburgh, for instance, at the beginning of the nineteenth century, the rising middle class was chafing under the social superiority of high society[18]. Applying the above strategy, they looked for unexpected allies to reverse this situation. They seized on a movement in brain science called phrenology that allowed almost anyone to read off people's qualities by carefully considering the bumps on their skulls and the shape of their faces. This use of cranial characteristics threatened to reshuffle Scottish class fabric entirely, exactly like the hygienists did above with the microbes (p.115). To evaluate the moral worth of someone the questions were no longer: Who are his parents? How ancient is his lineage? How vast are his propeties? But only: Does his skull possess the shape that expresses virtue and honesty? By allying themselves with phrenology the middle class could change its position in relation to the upper class, which at first was uninterested in brain science, by reallocating everybody into newly relevant groups. To resist brain scientists, *other* brain scientists had to be enlisted hook, line and sinker. Thus a controversy started not about social classes, but about neurology. As the controversy heated, the discussion shifted *inside* brain science; in fact, it shifted literally *inside* the brain. Atlases were printed, skulls cut open, dissections performed, to decide whether the inner structure of the brain could be predicted from the outer shape of the skull, as argued by phrenologists. Like the dissenters in Chapter 2 the newly recruited brain scientists tried out the connections established by phrenologists. The more they tried, the deeper they were led inside the brain, straining their eyes to discern whether the cerebellum, for instance, was linked to the rest of the body from the top or from the bottom. Moving slowly through the various translations, the contenders ended up in the cerebellum; and they did so because this latter proved the weak link.

(3) Machinations of forces

Interested groups may therefore be kept in line as, moving through a series of translations, they end up being trapped by a completely new element that is itself so strongly tied that nothing can break it up. Without exactly understanding how it all happened, people start placing transcontinental phone calls, taking photographs, having their cats and children vaccinated, and believing in phrenology. The quandary of the fact-builder is thus resolved, since all these people willingly contribute to the further expansion of these many black boxes. A new and deeper problem arises, however, caused by the very success of all the plots discussed above. These new and unexpected allies brought in to keep the first groups in line, *how can they*, in turn, *be kept in line?* Are they not another provisional juxtaposition of helping hands, ready to disband? Is not the flask of Pasteurian vaccine likely to be spoiled? What keeps the new prototype triodes from switching off after a few hours? What if the cerebellum turns out to be a shapeless mash of brain tissue? As to the diesel engine, we know how unreliable it is; it has to be debugged for longer years than the *Eagle* computer. How should these disordered assemblies be turned into such a tightly glued whole that it can link the enrolled groups together durably? Machiavelli knew perfectly well that the alliances binding towns and crowns are shifting and uncertain. But we are considering much more shifting and uncertain alliances between brains, microbes, electrons and fuels, than those necessary to bind together towns and crowns. If there is no way to render the new allies more reliable than the older ones, then the whole enterprise is spoiled and claims will shrink back to a single place and a single time.

We take the answer so much for granted that we no longer feel how simple and original it is. The simplest means of transforming the juxtaposed set of allies into a whole that acts as one is to tie the assembled forces *to one another*, that is, to build a **machine**. A machine, as its name implies, is first of all, a machination, a stratagem, a kind of cunning, where borrowed forces keep one another in check so that none can fly apart from the group. This makes a machine different from a tool which is a single element held *directly* in the hand of a man or a woman[19]. Useful as tools are, they never turn Mr of Ms Anybody into Mr or Ms Manybodies! The trick is to sever the link each tool has with each body and tie them to one another instead. The pestle is a tool in the woman's hand; she is stronger with it than with her hands alone, for now she is able to grind corn. However if you tie the grinder to a wooden frame and if this frame is tied to the sails of a mill that profits from the wind, this is a machine, a windmill, that puts into the miller's hands an assembly of forces no human could ever match.

It is essential to note that the skills required to go from the pestle to the windmill are exactly *symmetrical* to the ones we saw in Part A. How can the wind be borrowed? How can it be made to have a bearing on corn and bread? How can its force be translated so that, whatever it does or does not do, the corn is reliably ground? Yes, we may use the words translation and interest as well, because it is no more and no less difficult to interest a group in the fabrication of a vaccine than to interest the wind in the fabrication of bread. Complicated negotiations have to go on continuously in both cases so that the provisional alliances do not break off.

For instance, the assembled groups of farmers may, as I showed, lose interest. And the wind, what can it do? Simply blow the fragile windmill away, tearing the sails and the wings off. What should the mechanic do to hold the wind in his system of alliances, in spite of the way it shifts direction and changes strength? He has to negotiate. He has to tailor a machine that can stay open to the wind and still be immune to its deleterious effects. Severing the association between the sail mechanism and the tower on which the mill is built will do the trick. The top of the mill now revolves. Of course, there is a price to pay, for now you need more cranks and a complicated system of wheels, but the wind has been made into a reliable ally. No matter how much the winds shift, no matter what the winds want, the whole windmill will act as *one piece,* resisting dissociation in spite of/because of the increasing number of pieces it is now made of. What happens to the people gathered round the miller? They too are definitely 'interested' in the mill. No matter what they want, no matter how good they were at handling the pestle, they now have to pass through the mill. Thus they are kept in line *just as much* as the wind is[20]. If the wind had toppled the mill, then they could have abandoned the miller and gone their usual ways. Now that the top of the mill revolves, thanks to a complicated assembly of nuts and bolts, they cannot compete with it. It is a clever machination, isn't it, and *because* of it the mill has become an obligatory passage point for the people, for the corn and for the wind. If revolving windmills cannot do the job alone, then one can make it illegal to grind corn at home. If the new law does not work immediately, use fashion or taste, anything that will *habituate* people to the mill and forget their pestles. I told you the alliances were 'machiavellian'!

Still it is hard to see how a profusion of forces can be kept in line by relatively simple machinations like windmills. One snag becomes obvious: the process of recruiting and maintaining allies involves increasing complexity in the machine. Even the best mechanic will find it difficult to regulate the machine – check the wind, mend the sails, enforce the law – so that all the allies stay content. When you get to more complex machines, it's just a question of who/what breaks down first.

It would be better if the assembled forces could *check one another* by playing the role of mechanic for each other; if this were feasible, then the mechanic could withdraw and still benefit from the collective work of all the assembled elements, each conspiring with one another to fulfil the mechanic's goal. This would mean that, in practice, the assembled forces *would move by themselves!* This at first seems ludicrous, since it would mean that non-human elements would play the role of inspector, surveyor, checker, analyst and reporter in order to keep the assembled forces in line. It would mean another confusion of boundaries, the extension of social ploys to nature.

We are again so used to accepting the solution, that is hard for us to imagine how original the stratagems that generated **automatons** were. For instance, in the earlier Newcomen steam engine, the piston followed the condensing steam, pushed by atmospheric pressure, that was thus made to lend its strength to the pump that extracted the water, that flooded the coal mine, that made the pit useless...[21] A long series of associations, like those discussed in Part A, were made that linked the fate of coal mines to the weight of the atmosphere through the steam engine. The point here is that, when it reached the end of the cylinder, a new flow of steam had to be injected through a valve opened by a worker who then closed it again when the piston reached the top of its stroke. But why leave the opening and closing of the valve to a weary, underpaid and unreliable worker, when the piston moves up and down and could be *made to tell* the valve when to open and when to close? The mechanic who linked the piston with a cam to the valve transformed the piston into its own inspector – the story is that he was a tired, lazy boy. The piston is more reliable than the boy since it is, via the cam, *directly interested,* so to speak, in the right timing of the flow of steam. Certainly, it is more directly interested than any human being. An automatism is born, one of the first in a long series.

The engineer's ability lies in multiplying the tricks that make each element interested in the working of the others. These elements may be freely chosen among human or non-human actors[22]. For instance, in the early British cotton-spinning industry, a worker was attached to the machine in such a way that any failure of attention resulted not in a small deficiency in the product that could be hidden, but in a gross and obvious disruption which led to a loss of piecework earnings. In this case, it is part of the machine that is used to supervise the worker. A system of pay, detection of error, a worker, a cotton-spinning machine, were all tied together in order to transform the whole lash-up into a smoothly running automaton. The assembly of disorderly and unreliable allies is thus slowly turned into something that closely resembles an organised whole. When such a cohesion is obtained we at last have *a black box.*

Up to now I have used this term both too much and too loosely to mean either a well-establised fact or an unproblematic object. I could not define it properly before we had seen the final machinations that turn a gathering of forces into a whole that then may be used to control the behaviour of the enrolled groups. Until it can be made into an automaton, the elements that the fact-builder want to spread in time and space is not a black box. It does not *act as one.* It can be disassociated, dismantled, renegotiated, reappropriated. The Kodak camera is made of bits and pieces, of wood, of steel, of coating, of celluloid. The semi-professionals of the time open up their camera and do their own coating and developing, they manufacture their own paper. The object is dismembered each time a new photograph is taken, so that it is not one but rather a bunch of disconnected resources that others may plunder. Now the new Kodak automatic cannot be opened without going wrong. It is made up of many *more* parts and it is handled by a much *more* complex commercial network, but it acts as one piece. For the newly convinced user it is one object, no matter how many pieces there are in it and no matter how complex the commercial system of the Eastman Company is. So it is not simply a question of the number of allies but of their acting as a unified whole. With automatism, a large number of elements is made to act as one, and Eastman benefits from the whole assembly. When many elements are made to act as one, this is what I will now call a black box.

It is now understandable why, since the beginning of this book, no distinction has been made between what is called a 'scientific' fact and what is called a 'technical' object or artefact. This division, although traditional and convenient, artificially cuts through the question of how to ally oneself to resist controversies. The problem of the builder of 'fact' is the same as that of the builder of 'objects': how to convince others, how to control their behaviour, how to gather sufficient resources in one place, how to have the claim or the object spread out in time and space. In both cases, it is others who have the power to transform the claim or the object into a durable whole. Indeed, as we saw previously (Chapter 2) each time a fact starts to be undisputed it is fed back to the other laboratories as fast as possible. But the only way for new undisputed facts to be fed back, the only way for a whole stable field of science to be moblised in other fields, is for it to be turned into an automaton, a machine, one more piece of equipment in a lab, another black box. Technics and sciences are so much the same phenomenon that I was right to use the same term black box, even loosely, to designate their outcome.

Yet, despite this impossibility of distinguishing between science and technics, it is still possible to detect, in the process of enrolling allies and controlling their behaviour, two *moments* that will allow the reader to remain closer to common sense by retaining some difference between 'science' and 'technology'. The first moment is when new and unexpected allies are recruited – and this is most often visible in laboratories, in scientific and technical literature, in heated discussions;

the second moment is when all the gathered resources are made to act as one unbreakable whole – and this is more often visible in engines, machines and pieces of hardware. This is the only distinction that may be drawn between 'sciences' and 'technics' if we want to shadow scientists and engineers as they build their subtle and versatile alliances.

Part C
The model of diffusion
versus the model of translation

The task of the fact-builders is now clearly outlined: there is a set of strategies to enlist and interest the human actors, and a second set to enlist and interest the non-human actors so as to hold the first. When these strategies are successful the fact which has been built becomes indispensable; it is an obligatory passage point for everyone if they want to pursue their interests. From a few helpless people occupying a few weak points they end up controlling strongholds. Everyone happily borrows the claims or the prototypes from the successful contenders' hands. As a result, claims become well-established facts and prototypes are turned into routinely used pieces of equipment. Since the claim is believed by one more person, the product bought by one more customer, the argument incorporated in one more article or textbook, the black box encapsulated in one more engine, they spread in time and space.

If everything goes well it begins to look as if the black boxes were effortlessly gliding through space as a result of their own impetus, that they were becoming durable by their own inner strength. In the end, if everything goes really well, it seems as if there are facts and machines spreading through minds, factories and households, slowed down only in a handful of far-flung countries and by a few dimwits. Success in building black boxes has the strange consequence of generating these UFOs: the 'irreversible progress of science', the 'irresistible power of technology', more mysterious than flying saucers floating without energy through space and lasting for ever without ageing or decaying! Is this a strange consequence? Not for us since, in each chapter, we have learned to recognise the yawning gap that separates ready made science from science in the making. Once more, our old friend Janus is talking two languages at once: the right side is speaking in terms of *translations* about still undecided controversies, while the left side speaks of established facts and machines with the language of *diffusion*. If we want to benefit from our travels through the construction sites of science, it is crucial for us to distinguish between the two voices.

(1) Vis inertia . . .

In our examples we observed that the chain of people who borrowed claims varied from time to time because of the many elements the claims were tied to. If people wished to open the boxes, to renegotiate the facts, to appropriate them, masses of allies arrayed in tiers would come to the rescue of the claims and force the dissenters into assent; but the allies will not even think of disputing the claims, since this would be against their own interests which the new objects have so neatly translated. Dissent has been made unthinkable. At this point, these people do not do anything more to the objects, except pass them along, reproduce them, buy them, believe them. The result of such smooth borrowing is that there are simply more copies of the same object. This is what happened to the double-helix after 1952, to the *Eclipse MV/8000* after 1982, to Diesel's engine after 1914, to the Curies' polonium after 1900, to Pasteur's vaccine after 1881, to Guillemin's GRF after 1982. So many people accept them that they seem to flow as effortlessly as the voice of Alexander Bell through the thousands of miles of the new transcontinental line, even though his voice is amplified every thirteen miles and completely broken down and recomposed six times over! It also seems that all the work is now over. Spewed out by a few centres and laboratories, new things and beliefs are emerging, free floating through minds and hands, populating the world with replicas of themselves.

I will call this description of moving facts and machines the **diffusion model**. It has a number of strange characteristics which, if taken seriously, make the argument of this book exceedingly difficult to grasp.

First, it seems that as people so easily agree to transmit the object, it is the object itself that forces them to assent. It then seems that the behaviour of people is *caused* by the diffusion of facts and machines. It is forgotten that the obedient behaviour of people is what turns the claims into facts and machines; the careful strategies that give the object the contours that will provide assent are also forgotten. Cutting through the many machiavellian strategies of this chapter, the model of diffusion invents a technical determinism, paralleled by a scientific determinism. Diesel's engine leaps with its own strength at the consumer's throat, irresistibly forcing itself into trucks and submarines, and as to the Curies' polonium, it freely pollinates the open minds of the academic world. Facts now have a *vis inertia* of their own. They seem to move even without people. More fantastic, it seems they would have existed even without people at all.

The second consequence is as bizarre as the first. Since facts are now endowed with an inertia that does not depend on the action of people or on that of their many non-human allies, what propels them? To solve this question adepts of the diffusion model have to invent a new mating system. Facts are supposed to reproduce one another! Forgotten are the many people who carry them from hand to hand, the crowds of acting entities that shape the facts and are shaped by them, the complex negotiations to decide which association is stronger or weaker; forgotten are the three chapters above, as from now on we reach the realm of ideas begetting ideas begetting ideas. Despite the fact that it is hard to picture Diesel's engines or bicycles or atomic plants reproducing themselves through mating, trajectories (see p.107) are drawn that look like lineages and genealogies of 'purely technical' descent. The history of ideas, or the conceptual history of science, or epistemology, these are the names of the discipline – that often should be X-rated – that explains the obscure reproduction habits of these pure breeds.

The problem with the mating system of facts that diffuse through their own force is novelty. Facts and machines are constantly changing and are not simply reproduced. Nobody shapes science and technologies except at the beginning, so, in the diffusion model, the only reasonable explanation of novelty lies with the initiators, the first men and women of science. Thus, in order to reconcile inertia and novelty the notion of *discovery* has been invented; what was there all along (microbes, electrons, Diesel's engine) needs a few people, not to shape it, but to help it to appear in public.[23] This new bizarre 'sexual reproduction' is made half by a history of ideas and half by a history of great inventors and discoverers, the Diesels, the Pasteurs, the Curies. But then there is a new problem. The initiators, in all the stories I have told, are only a few elements in a crowd. They cannot be the cause of such a general movement. In particular, they cannot be the cause of the people who believe them and are interested in their claims! Pasteur has not enough strength to propel his vaccine across the world, nor Diesel his engine, nor Eastman his Kodak. This is not a problem for our 'diffusionists'. They simply make the inventors so big that they now have the strength of giants with which to propel all these things! Blown out of proportion, great men and women of science are now geniuses of mythological size. What neither Pasteur nor Diesel could do, these new figures also named 'Pasteur' and 'Diesel' can. With their fabulous strength it is a cinch for these Supermen to make facts hard and machines efficient!

Great initiators have become so important for the diffusion model that its advocates, taken in by their own maniac logic, have now to ferret out *who* really was the first. This quite secondary question becomes crucial here since *the winner takes all*. The question of how to allocate influence, priority and originality among great scientists is taken as seriously as that of discovering the legitimate heir of an empire! Labels of 'precursor', or 'unknown genius', or 'marginal figure', or 'catalyst', or 'driving force' are the object of punctilios as ornate as etiquette at Versailles at the time of Louis XIV; historians rush forward to provide genealogies and coats of arms. The secondary mechanism takes precedence over the primary mechanism.

The funniest thing about this fairy tale is that, no matter how carefully these labels are attributed, the great men and women of science are always a few names in a crowd that cannot be annihilated even by the most enthusiastic advocates of the diffusion model. Diesel, as we saw, did not make everything of the engine that bears his name. Pasteur is not the one that made asepsis a workable practice, or stopped millions of people from spitting, or distributed the doses of vaccine. Even the most fanatic diffusionists have to grant that. However this does not bother them. Going further and further into their fantasies, they invent geniuses who did it all, but only 'in the abstract', only 'seminally', only 'in theory'.

Sweeping away the crowds of actors, they now picture geniuses that *have ideas*. The rest, they argue, is mere development, a simple unfolding of the 'original principles' that really count. Thousands of people are at work, hundreds of thousands of new actors are mobilised in these works, but only a few are designated as the motors that move the whole thing. Since it is obvious that they did not do that much, they are endowed with 'seminal ideas'. Diesel 'had the idea' of his engine, Pasteur 'had the idea of asepsis' . . . It is ironic to see that the '**ideas**,' which are so valued when people talk of science and technology, are a trick to get away from the absurd consequences of the diffusion model, and to explain – away – how it is that the few people who did everything nevertheless did so little.

The model of diffusion would be rather quaint and insignificant if it were not for its final consequence which is taken seriously even by those who are willing to study the inner workings of technoscience.

Attentive readers who accept what we have argued so far might think it is easy to question the diffusion model. If the interpretation given by the model is ludicrous, the impression from which it springs is genuine. It seems to work in the few cases when facts and artefacts convince people and, for this reason, seem to flow. Thus, readers may think that the diffusion model will break apart when the facts are interrupted, deflected, ignored or corrupted. The action of many people will necessarily irrupt into the picture, since there is no one at hand to 'diffuse' the facts any more. Well, if they think so, it simply means that these readers are still naive and that they underestimate the ability of an interpretation to hold out against all contrary evidences. When a fact is not believed, when an innovation is not taken up, when a theory is put to a completely different use, the diffusion model simply says that 'some groups resist'.

In the story of Pasteur, for instance, adepts of the diffusion model have to admit that physicians were not very interested in his results; they thought that these were premature, unscientific, and of little use. Indeed, they did not have much use for vaccines since *preventive* medicine was taking business away from them. Instead of looking at how the research program of the Institut Pasteur was being constantly modified by dozens of people in order to convince almost every physician, the diffusion model simply says that Pasteur's ideas were *blocked by* certain groups which were stupid or had 'vested interests' in older techniques. They picture the physicians as corporatists, as selfish, as a backward and reactionary group, that slowed down the spread of Pasteur's idea for a generation. So the diffusion model traces a dotted line along the path that the 'idea' should have followed, and then, since the idea did not go very far and very fast, they make up groups that resist. With this last invention, both the principle of inertia and the fantastic force that triggers it at the beginning are maintained, and the gigantic stature of the great men and women that gave momentum to the whole is amplified. Diffusionists simply add *passive* social groups to the picture that may, because of their own inertia, slow down the path of the idea or absorb the impact of the technics. In other words, the diffusion model now invents *a society* to account for the uneven diffusion of ideas and machines. In this model, society is simply a medium of different resistances *through which* ideas and machines travel. For instance, the Diesel engine that has spread through the developed countries because of the momentum given to it by Diesel might slow down or even stop in some underdeveloped country where it rusts on a dock in the tropical rain. In the diffusion model, this would be accounted for in terms of the resistance, the passivity or the ignorance of the local culture. Society or 'social factors' would appear only at the end of the trajectory, when something went wrong. This has been called the principle of *asymmetry*: there is appeal to social factors only when the true path of reason has been 'distorted' but not when it goes straight.[24]

The society invented to maintain the diffusion model has another strange characteristic. The 'groups' that make it up do not always interrupt or deflect the normal and logical path of ideas; they may suddenly switch from being resistors or semiconductors to conductors. For instance, the same physicians who were not very happy with Pasteur until 1894 then became all of a sudden interested in the Pasteurians' work. This is not a difficulty in the diffusion model: they simply altered their position. They switched open. The resistors began to conduct, the reactionaries to progress, from being backward they suddenly moved forward! You see that there is no limit to the fairy tale. Forgotten is the careful co-production between Pasteurians and physicians of a new object, a serum against diphtheria that, unlike the preventive vaccine, was at last one that helped to *cure*. The long translations necessary to convince horses, diphtheria, hospitals and physicians to associate with one another in this new object are forgotten. Cutting across the complicated systems of associations, the diffusion model simply extracts a serum – that was there all along, at least 'in principle' – and then invents groups which at first resisted and finally 'turned out' to accept the discovery.

(2) Weaker and stronger associations

Let us go back to Diesel in order to understand the differences between the diffusion model and the translation model. We saw that Diesel's engine was a sketch in his patent, then a blueprint, then one prototype, then a few prototypes, then nothing, then again a single new prototype, then no longer a *proto*type but a *type* that was reproducible in several copies, then thousands of engines of different sub-types. So there was indeed a proliferation. First, following the translations, we learned that this increase in the number of copies had to be paid for by an increase in the number of people made to be interested in its fate. Second, we realised that this increase in copies and people had to be obtained through a deep transformation of the design and principles of the engine; the engine moved, but it was not the same engine. Third, we learnt that it had been transformed so much during the translation that there was a dispute about whose engine it actually was. And fourth, we saw that in about 1914 there had been a point when people could accept the engine not as a prototype but as a copy, and take it away from the Augsburg shop without deeply transforming it or dragging with them dozens of mechanics and patent lawyers; the engine was a black box for sale at last and it was able to interest not only engineers and researchers but also 'simple customers'. It is at this point that we left the story, but it is also at this point that the diffusion model seems better than the translation one because no one is necessary any more to shape the black box. There exist only customers who buy it.

How simple is a 'simple customer'? The customer is 'simple' because he or she does not have to redesign the engine by shifting from air injection back to solid injection, or moving the valves around, or boring new cylinders and running the engine on the test bench. But the customer cannot be so 'simple' as not to tend the engine, feeding it oil and fuel, cooling it, overhauling it regularly. Even when the phases of development and innovation have ended, the darkest black box still has to be *maintained* in existence by not so simple customers. We can easily picture endless situations in which an ill-informed or a stupid consumer makes one engine falter, or stall or blow apart. As engineers say, no device is idiot-proof. This particular copy of the engine at least will not run any more, but will slowly rust.

There is another problem with 'simple' customers. Let us remember Eastman's Kodak camera. It was simpler to operate than anything before. 'Push the button, we'll do the rest,' they said. But they had to do *the rest*, and that was quite a lot. The simplification of the camera that made it possible to interest everyone in its dissemination in millions of copies had to be obtained by the extension and complication of Eastman's commercial network. When you push the button you do not see the salesmen and the machines that make the long strips of celluloid films and the trouble-shooters that make the coating stick properly at last; you do not see them, but they have to be there none the less. If they are not, you push the button and nothing happens. The more automatic and the blacker the black box is, the more it has to be *accompanied* by people. In many situations, as we all know all too well, the black box stops pitifully because there is no salesperson, no repairer, no spare part. Every reader who has lived in an underdeveloped country or used a newly developed machine will know how to evaluate the hitherto unknown number of people necessary to make the simplest device work! So, in the most favourable cases, even when it is a routine piece of equipment, the black box requires an active customer and needs to be accompanied by other people if it is to be maintained in existence. By itself it has no *inertia*.

If we have understood this, then we may draw the conclusions from the two first parts of this chapter: the black box moves in space and becomes durable in time only through the actions of many people; if there is no one to take it up, it stops and falls apart however many people may have taken it up for however long before. But the type, the number and the qualifications of the people in the chain will be modified: inventors like Diesel or Eastman, engineers, mechanics, salesmen, and maybe 'ignorant consumers' in the end. To sum up, *there are*

always people moving the objects along but they are not the same people all along.
Why are they not the same? Because the first ones have tied the engine's fate to
other elements so that the engine may be put in different hands and more easily
spread. You will then see a few copies of the Diesel engine slowly move through
its constant redesign at the test bench, and suddenly you will observe many copies
of the same design that are bought and sold by many people. There are always
people, but they are not the same. Thus, the diesel engine story may be analysed
either by looking at the changing shape of the engine – tied to different
people – or by looking at the changing type of people – linked to the engine. It is
the *same story* viewed either from the standpoint of the enrolled people of Part A
or from the enrolling things of Part B.

Similarly, the Curies' polonium was first a claim redesigned after every trial in
a single laboratory in Paris in 1898. To convince dissenters that this was indeed a
new substance, the Curies had to modify the trials and renegotiate the definition
of their object. For each suspicion that it might be an artefact, they devised a trial
that linked its fate to a more remote and less disputable part of physics. There is a
moment in this story when the claim becomes a new object, and even a part of
Nature. At this point the type of people necessary to provide the fact with
durability and extension is to be modified. Polonium may now travel from the
Curies' hands into many more, but much less informed, hands. It is now a routine
radioactive element in a sturdy lead container, one more box filled up in freshly
printed versions of the periodic table; it is no longer believed by only a few bright
sparks in a few laboratories, but also by hundreds of enthusiastic physicists; soon
it will be learned by 'simple students'. A continuous chain of people using, testing
and believing in polonium is necessary to maintain it in existence; but they are not
the same people nor are their qualifications the same. So the story of
polonium – like all that have so far been told in this book – may be told either by
looking at the people who are convinced, or by looking at the new associations
made to convince them. It is the same analysis from two different angles since, all
along, polonium is constituted by *these* people convinced that *these* associations
are unbreakable.

We may now generalise a bit from what we have learned. If you take any black
box and make a freeze-frame of it, you may consider the system of alliances it
knits together in two different ways: first, by looking at who it is designed to
enrol; second, by considering what it is tied to so as to make the enrolment
inescapable. We may on the one hand draw its **sociogram**, and on the other its
technogram. Every piece of information you obtain on one system is also
information on the other. If you tell me that Diesel's engine now has a stable
shape, I will tell you how many people at MAN had to work on it and about the
new system of solid injection they had to devise so that the engine might be
bought by 'mere consumers'. If you tell me that you think polonium is really
bismuth (see p.88), I can tell you that you work in the Curies' lab in Paris around
1900. If you show me a serum for diphtheria, I'll understand how far you drifted
from the original research programme that aimed at making vaccines and I'll tell

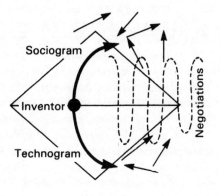

Figure 3.4

you who are the physicians who will get interested. If you show me an electric vehicle running on fuel cells, I'll know who has to be won over in the company. If you propose to build a 16-bit computer to compete with the DEC's *VAX 11/780* machine I'll know who, when and where you are. You are West at Data General in the late 1970s. I know this, because there are very few places on earth where anyone has the resources and the guts to disaggregate the black box DEC has assembled and to come up with a brand new make of computer. I similarly learn a lot about you if you explain to me that you are waiting for the repair man to fix your Apple computer, or that you believe the moon to be made of green cheese, or that you do not really think that the second amino acid in the GHRH structure is histidine.

Carefully take note that the black box is *in between* these two systems of alliances, that it is the obligatory passage point that holds the two together and that, when it is successful, it concentrates in itself the largest number of hardest associations, especially if it has been turned into an automaton. This is why we call such black boxes 'hard facts', or 'highly sophisticated machines', or 'powerful theories', or 'indisputable evidence'. All these adjectives that allude to strength and power rightly point out the disproportionate number of associations gathered in these black boxes, so disproportionate indeed that they are what keep the multitude of allies in place. However this disproportion often leads us to forget that they hold things and people tightly together only as long as all the other strategies are successful. Do these products of science and technics escape from the system of complicated alliances with which politics are managed, for instance? Are they less 'social' as people often naively say? Most unlikely; if they had to be qualified in these terms – which they don't – they would have to be described as *more*, much more 'social'.

If you now let the frozen-frame move, you observe a black box that simultaneously changes what it is made of and whom it is convincing. Each modification in one system of alliances is visible in the other. Each alteration in the technogram is made to overcome a limitation in the sociogram, or vice versa. Everything happens as if the people we have to follow were in between two sets of constraints and were *appealing* from one to the other whenever the negotiations get stalled. On one side there are people who are either going in the same direction, or are against it, or are indifferent, or, although indifferent and hostile, may be convinced to change their minds. On the other side, there are non-human actors in all colours and shades: some are hostile, others indifferent, some are already docile and convenient, still others, although hostile or useless, may be persuaded to follow another path. The inventor of Post-it, a yellow sticky paper for marking books, which has now become so widely used, makes the point very well.[25] Having found a glue that does *not* adhere was seen as a failure in the 3-M company whose job is usually to make very sticky glues. This failure to glue was turned to advantage when the inventor realised that it could mark Psalm books without smearing or wearing them. Unfortunately, this advantage was not admitted by the marketing department who had decided that this invention had no market and no future. Situated exactly at the middle of the techno- and of the sociograms, the inventor has a choice: either to modify the invention or to modify the marketing department. Choosing to keep the invention as it is, he then applies subtle tactics to sway the marketing department, distributing prototypes of his invention to all the secretaries, and then asking the secretaries, when they wanted more of it, to call the marketing department directly! It is the same subtlety that goes on in devising a glue that does not glue or in making a marketing department sell what they do not want to sell. Rather, Post-it is shaped by the two sets of strategies, one for enrolling others, the other to control their behaviour.

We may go a bit further. We are all multi-conductors and we can either drop, transfer, deflect, modify, ignore, corrupt or appropriate the claims that need our help if they are to spread and last. When – very rarely – the multi-conductors, acting as conductors, simply transmit a belief without delay and corruption, what does this mean? That many elements accompany the moving claims or objects and literally *keep* the successive hands necessary for their survival *in line*. When – more often – multi-conductors *interrupt* the spread of the claims that had until then been passed along without qualms by everyone, it also teaches us something. Since they are able to interrupt, these people must be tied to new interests and new resources that counteract the others. And the same lessons may be drawn when – as is almost always the case – people ignore, deflect, modify or appropriate the black boxes. Does the reader now see the conclusion? *Understanding* what facts and machines are is the same task as understanding who the people are. If you describe the controlling elements that have been gathered together you will understand the groups which are controlled. Conversely, if you observe the new groups which are tied together, you will see how machines work and why facts are hard. The only question in common is to learn *which associations are stronger and which weaker*. We are never confronted with science, technology and society, but with a gamut of weaker and stronger associations; thus understanding *what* facts and machines are is the same task as understanding *who* the people are. This esssential tenet will constitute our **third principle**.

(3) The fourth rule of method

Among all the features that differ in the two models, one is especially important, that is society. In the diffusion model society is made of groups which have interests; these groups resist, accept or ignore both facts and machines, which have their own inertia. In consequence we have science and technics on the one hand, and a society on the other. In the translation model, however, no such distinction exists since there are only heterogeneous chains of associations that, from time to time, create obligatory passage points. Let us go further: *belief* in the existence of a society separated from technoscience *is an outcome of the diffusion model.* Once facts and machines have been endowed with their own inertia, and once the collective action of human and non-human actors tied together has been forgotten or pushed aside, then you have to make up a society to explain why facts and machines do not spread. An artificial division is set up between the weaker and stronger associations: facts are tied with facts; machines with machines; social factors with social factors. This is how you end up with the idea that there are three spheres of Science, Technology and Society, where the influence and impact of each on the other have to be studied!

But worse is yet to come. Now that a society has been invented by artificially cutting through the associations and the translations, and by squeezing social factors into tiny ghettos, some people try to explain science and technology by the influence of these social factors! A social or a cultural or an economic determinism is now added to the technical determinism above. This is the meaning of the word social in expressions like 'social studies of science' or 'the social construction of technology'. Analysts who use groups endowed with interests in order to explain how an idea spreads, a theory is accepted, or a machine rejected, are not aware that the very groups, the very interests that they use as *causes* in their explanations are the *consequence* of an artificial extraction and purification of a handful of links from these ideas, theories or machines. Social determinism courageously fights against technical determinism, whereas *neither exist* except in the fanciful description proposed by the diffusion model.

Although there is no point in spending too much time on the diffusion model it is crucial, if we wish to continue our voyage through technoscience, to be immunised against the notion that there is a society and 'social factors' able to shape, influence, direct or slow down the path of pure science and pure technics. At the end of Chapter 2, I presented our third rule of method: Nature cannot be used to account for the settlement of controversies, because it is only after the controversies have been settled that we know what side she is on. 'Nature settles only the settled claims,' so speaks the left side of our Janus who does not sense the contradiction. As for the unsettled ones on which the right side of Janus is

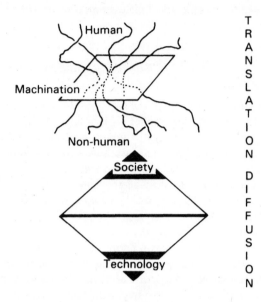

Figure 3.5

working, we do not yet know what settles them but it is not Nature. Nature thus lies behind the facts once they are made; never behind facts in the making.

If we want to go on without being bothered by the diffusion model, we have to offer a fourth rule of method, as basic to the third one, and symmetrical to it, which applies this time *to society*.

Right from the first pages of this book the reader may have noticed the shocking absence of the entities that traditionally make up Society, an absence that may be even more shocking than the delayed appearance of Nature until the end of Chapter 2. After three chapters there has been not a word yet on social classes, on capitalism, on economic infrastructure, on big business, on gender, not a single discussion of culture, not even an allusion to the social impact of technology. This is not my fault. I suggested that we follow scientists and engineers at work and it turns out that *they do not know what society is made of*, any more than they know the nature of Nature beforehand. It is because they know about neither that they are so busy *trying out* new associations, creating an inside world in which to work, displacing interests, negotiating facts, reshuffling groups and recruiting new allies.

In their research work, they are never quite sure which association is going to hold and which one will give way. Diesel was confident at first that all fuels would ignite at high temperatures and that every group of users would be interested in his more efficient engine. But most fuels rejected his engine and most consumers lost interest. Starting from a stable state of Nature and of Society, he had to struggle through another engine tying kerosene, air injection and a tiny number of users together. Hygienists also started with a fixed state of Society – the class struggle – and a determined state of Nature – the miasmatic diseases. When Pasteurians offered them the microbes, this was a new and unpredictable definition both of Nature and of Society: a new social link, the microbe, tied men and animals together, and tied them differently. There was nothing in the stable state of either Society or Nature that made an alliance of big business at Bell with electrons necessary or predictable. The Bell Company was deeply modified by its alliance with Millikan's physics, it was not the *same* Bell, but neither was it the same physics, the same Millikan nor, indeed, the same electrons. The versatility and the heterogeneity of the alliances is precisely what makes it possible for the researchers to get over the quandary of the fact-builder: how to interest people and to control their behaviour. When we study scientists and engineers at work, the only two questions that should not be raised are: What is Nature really like? What is Society really made of?

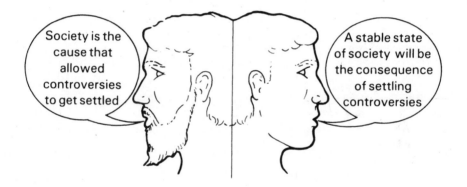

Figure 3.6

To raise these questions we have to wait until scientists and their allies – among whom social scientists should of course be included – have finished their work! Once the controversies have ended, then a stable state of Society, together with a stable rendering of the interests of its members, will emerge. If we study all made facts and groups, then interests and Nature will be clearly articulated by the left face of Janus. Not so, when we follow facts in the making. It might seem a strange consequence but it is a necessary one: to follow scientists and engineers we do not need to know what Society is made of and what Nature is; more exactly, we need *not* to know them. The stable state of Society is three chapters away! The premature introduction of a fully-fledged Society would be as damaging for our trip as would a complete picture of Nature. More exactly the same arguments that have been made about Nature have to be made *symmetrically* about Society. How could we take so many precautions in *not* believing directly what scientists and engineers say about objectivity and subjectivity, and readily believe what other scientists (social this time) say about society, culture and economy? At this point we are in great need of a rule of symmetry that does not grant Society privileges refused to Nature. Our **fourth rule of method** thus reads exactly like the third – the word 'Society' replacing the word 'Nature' – and then fuses the two together: since the settlement of a controversy is the *cause* of Society's stability, we cannot use Society to explain how and why a controversy has been settled. We should consider *symmetrically* the efforts to enrol and control human and non-human resources.

CHAPTER 4

Insiders Out

We now have a better idea of the amount of preliminary work necessary to secure enough strongholds to make relevant the added force offered by the technical literature and the laboratories. Without the enrolment of many other people, without the subtle tactics that symmetrically adjust human and non-human resources, the rhetoric of science is powerless. People escape, lose interest, do something else, are indifferent. Still, the stories told in the former chapter were all from the point of view of the enlisting scientists and engineers. Even if we had followed many more outcomes than the three we started with – giving up, going along, working through – we might have had the impression that scientists and engineers were at the *centre* of everything. This impression might create some new difficulties. Our first rule of method requires us to *shadow* scientists while they are engaged in their work of doing science. At face value this precept seems easy to put into practice; this is why, in all the chapters so far, I have pretended that we at least knew *where* to find the white-coated protagonist to begin our enquiry. But it was to simplify our trip that I took it for granted that West, Crick and Watson, Guillemin, the Professor, Diesel, Mead or Pasteur were able to gather resources, to talk with authority, to convince others of their strength and to equip laboratories or departments, thus beginning the various stories I told with fully-fledged scientists and engineers that others were taking seriously enough to grant them attention, money and confidence. To offer us a convenient departure point I invented a character whom I called the 'dissenter' to help us practise the difficult art of shadowing scientists in action; and indeed, since this dissenter was easy to detect and since his obstinacy made him easier to follow, it facilitated our peregrination through the technical literature and through laboratories. Later, the character of the 'fact-builder' was very convenient to map the various types of translations.

Nothing proves, however, that following real scientists and engineers is as easy as following these dummy dissenters or dummy fact-builders, especially when the very principles we uncovered hint at the opposite. Remember that the first basic principle states that facts are made collectively, the second that scientists and engineers speak in the name of new objects shaped by unexpected trials of stength and the third that describing facts and machines is the same thing as describing the people they enrol and control. Many new questions arise from these principles: since there is not much difference between those who enlist and those who are enlisted why should we concentrate on scientists? Who are the people collectively working in fact-construction? Are they all scientists and engineers? If they are not, what the hell *are* they doing? If scientists are spokespersons, to whom are they speaking? Who are the other representatives? How do they settle their controversies?

When raising these questions we begin to realise that it might not be so easy to determine who are the scientists and the engineers, and thus to decide, as is required by our first rule of method, *whom* to follow. We have no choice, however, and we should stick as stubbornly as ever to our task, with the addition of more subtlety now that our guide is going to wear many confusing masks and to follow multifarious paths simultaneously.

Part A
Interesting others
in the laboratories

(1) When everyone can do without scientists or engineers

What happens to scientists and engineers who have not secured any strongholds? How strong will their rhetoric be? How capable will they be of keeping interest groups in line? Let me take two examples, one of a scientist in the past and one of an engineer in the present. In these examples no one is prepared to grant anything to the budding researchers and everyone does very well *without* their science.

(A) WHEN BEING A SCIENTIST IS NOT YET A JOB

In the late 1820s, Charles Lyell was reading for the Bar and living on a £400-a-year allowance from his upper-middle-class father[1]. Lyell wished to study the 'history of the earth'. Do not jump to the conclusion that he wanted to be a geologist. Being able to be a geologist will be the *result* of the work of many people like Lyell. At the time there was no such thing in England as a full-time paid and secure job under the label 'geologist'. Moreover, 'geology' did not really exist either. The history of the earth pertained to theology and biblical exegesis as well as to paleontology and other technical subjects. In other words, neither the discipline of geology nor the profession of geologist existed. One of the related and firmly established disciplines was that of the 'rational history of creation' and one of the related trades is a six-century-old profession, that of cleric in the universities – with compulsory celibacy, at least at Cambridge.

When he starts, there is no laboratory which Lyell can enter, no curriculum to follow and no grant for which to apply. Although Lyell needs others to help him build new and harder facts these 'others' are following different tracks. Can Lyell count on the dons and clerics of Oxford who teach the history of the earth and who have the libraries, the authority and the tenures? Not at all, because, if a controversy is triggered about, say, the age of the earth, Lyell's colleagues may very well *interrupt* his argument by appealing to God's word or to the Church's perennial teachings. Even if the dons Lyell is addressing are interested in a rational history of the earth and have agreed to talk about rocks and erosion without bringing in the location of the Garden of Eden, the size of Noah's Ark or the date of the Flood, what will happen if the controversy heats up a bit? Not much, simply because these colleagues have taken the chair as a first step toward becoming either bishop or teacher of a more prestigious subject, like ethics. No matter how many arguments Lyell has been able to muster in defence of his position, his opponents are in no way forced to take up his point. They may simply ignore him, or brush the arguments aside, or listen with bewilderment and go on teaching their usual course. For the dissenter to exist more work has to be done.

The same thing might happen if Lyell sets up a controversy with the miscellaneous groups of people who write 'theories of the earth' on the side, but who do not make a living from geology, that is the *amateurs*. Many amateurs were busy at the time gathering rocks and fossils, visiting foreign landscapes, offering all sorts of reports to the many societies recently created to gather new collections. By definition, an amateur, even a devoted and a passionate one, may leave the discussion whenever it pleases him. So it is very hard for Lyell to win an argument and to force the amateur to borrow his claims as a black box, especially if they run against his feelings, interests and passion. Unconvinced, the amateurs may go on as usual, uninterested and unthreatened by the many allies that Lyell may have assembled in support of his position. Although they are necessary to collect the rocks and the fossils in many places where the few geologists could not possibly go, the amateurs form a most undisciplined crowd as far as helping Lyell produce new facts goes.

The situation would be much better for Lyell if the clerics would give up their chairs in universities and hand them over to people with no other ambitions than to stay *inside* geology all their lives. Geology would then become a career. When Lyell makes a point, his colleagues would have to either defeat him or accept it because they would have no other way to go. They could no longer ignore him or do something else such as becoming a bishop. It would also be better if the amateurs were still busy gathering materials and providing reports, but were not meddling in the debates. They would be forced to bring in their specimens, to offer their collections, but they would stay *outside* without adding their own commentaries and theories. A disordered crowd of helping hands would then become a disciplined workforce helping geologists produce more documented facts. Slowly an inside pocket of purely geological matters would be carved out of the outside world, and the author-dissenter duel of Chapters 1 and 2 could take place.

The problem is that even if Lyell had succeeded in creating an assembly of colleagues who did nothing else but geology, none of them would be able to secure a salary or at least to offer him one. So Lyell has to earn a living elsewhere, his father's pittance not being enough to raise a family and to gather a collection. Since he is a bright lecturer and likes the leisurely life of the upper class, one solution is to address the enlightened gentry. However, this leads him into new difficulties. First, he may dissipate his time in worldly circles explaining the mystery of the Precambrian rocks to the Earl of This and the Baroness of That. Even if he is successful and gathers a large audience of paying gentry, he might have no time left to produce new facts; hence, he will end up *teaching* geology as it is, not making geology anew. Lyell would indeed be outside collecting resources but he would never bring them back inside!

The situation would be worse still if, in order to make his teaching acceptable and understandable, he had to negotiate the very content of his lesson with the amiable but flighty and unprofessional assembly. For instance, his audience might be shocked by the age that Lyell gives to earth history, since they imagine they are living in a world a few thousand years old, whereas Lyell needs at least a several-million-year framework for his geology. If he lets the audience participate in the production of the facts, Lyell is faced with a new dilemma: make the earth younger so as not to lose his audience, or age the earth but be left with no one to attend his lectures! No, the ideal would be if the interested and literate audience could pay for geology, waiting *outside* for Lyell and his colleagues to develop it as they see fit, and then, later, would be allowed to learn what the age of the earth is without trying to negotiate the facts. Even this would not be sufficient, because these noblemen and women might be too frivolous to wait long enough for thousands of fossils to be gathered in numerous collections. Their interests might fade rapidly, replaced by a new fashion for electricity or magnetism or anthropology. No! for the situation to be ideal, money should flow regularly and irreversibly without depending on mood and fashion, something as compulsory and as regular as a *tax*.

To obtain such a result, Lyell would have to interest not only the gentry but high officials of the state, and to convince some agency that geology may be relevant and useful for their aims. As we saw in Chapter 3, Part A, this translation of interest is possible if geology is able to produce a great number of new and unexpected facts, which can then be seen as resources for some of the state's problems – finding new coal deposits, substituting strategic minerals to others, reclaiming new land, mapping new territories, and so on. However, the assembled interests can be held in place only if Lyell is able to speak in the name of many *new objects*, which supposes an *already* existing science. Conversely, the production of hard facts is impossible without the collective work of many full-time scientists and devoted amateurs digging up rocks, visiting rift valleys and canyons, surveying the land and bringing huge collections of rocks and fossils into Natural History Museums, as the French geologists were doing at the time in Paris.

At the beginning of this science, Lyell is in a vicious circle: an ill-funded geology will not interest the state and so will remain too weak to resist the competition of other disciplines and priorities. This is the opposite to the starting point of all our stories so far, in which everyone helps in the strengthening of the scientists' and engineers' laboratories. Instead of being welcomed by high officials, newsmen, priests, students and industrialists, Lyell may simply be ignored. Even if he tries, so to speak, to oversell the discipline *before* it has achieved results, he may run into a new danger. Organising the profession, imposing stringent standards on the training of young colleagues, promoting new ways of settling controversies, new journals, new museums, kicking the amateurs out, lobbying the state, advertising the future results of the disclipline, all that takes time, so much time that Lyell once more may not be able to contribute the reshaping of the earth which he is aiming at.

Of course, he could appeal to a larger public in writing, as he did for instance in his *Principles of Geology.* If this book were to become a best seller, then Lyell might have money to gather new resources and produce new facts. But this is running another risk. How should he appeal to the public? If his *Principles* are to interest everyone, then he might have to eliminate the technical details, but then he might become one of these amateurs, popularisers or pamphleteers of geology, no longer a geologist. But, if Lyell's book engages in controversies and reshapes everyone's belief by bringing in new resources, we know what will happen (Chapter 1); the book will become technical, so technical that there will be no one left to read it. Lyell will still be without money to further his research.

Even if Lyell is clever enough to solve this problem, then he may stumble over another one. If geology is successful in reshaping the earth's history, size, composition and age, by the same token, it is also extremely shocking and unusual. You start the book in a world created by God's will 6000 years ago, and you end it with a few poor Englishmen lost in the eons of time, preceded by hundreds of Floods and hundreds of thousands of different species. The shock might be so violent that the whole of England would be up in arms against geologists, bringing the whole discipline into disrepute. On the other hand, if Lyell softens the blow too much, then the book is not about new facts, but is a careful compromise between common sense and the geologists' opinion. This negotiation is all the more difficult if the new discipline runs not only against the Church's teachings but also against Lyell's own beliefs, as is the case with the advent of humanity into earth history which Lyell preferred to keep recent and miraculous despite his other theories. How is it possible to say simultaneously that it is useful for everyone, but runs against everyone's beliefs? How is it possible to convince the gentry and at the same time to destroy the authority of common sense? How is it possible to assert that it is morally necessary to develop geology while agonising in private in the meantime on the position of humanity in Nature?

It is not an easy job being a scientist before the job exists! Before others may set foot inside geology, Lyell has to fight outside on all fronts at once. He has to eliminate amateurs – but needs to retain them as a disciplined workforce, to please the gentry and gather their wealth – but to keep them at arm's length so as not to waste time and discuss their opinions; he has to prove to the state that geology is the most important thing on earth, an obligatory passage point for things they want to do and that, for this reason, they should provide well-paid jobs – but he should also delay their expectations, make their scrutiny impossible, avoid all state incursions and force them not to ask too much in exchange; he has to fight endlessly against the Church and the dons – but also to find a way to sneak geologists inside the old universities' curricula where tenures can be obtained; finally, he has to appeal to the multitude for support and enthusiasm – but he should do so without shocking them while shattering their world-view! Yes, there is one other thing he needs to do besides all that fighting: research in geology. It is only when the above battles have been partially successful that he may win colleagues over in the collective construction of some new arguments about the earth[2].

(B) A NON-OBLIGATORY PASSAGE POINT

Lyell had to create simultaneously the outside and the inside of geology. At the beginning everyone could do without him; at the end of the century, geology had become indispensable for many other sciences, professions, industries, and state ventures. Geologists at work a century after Lyell would look very much like the dissenters and fact-builders of the other chapters; like them, they would have to cater to others' interests. Although they would have to be clever and interesting, there would be no question about the basic importance of their discipline. Most of the groundwork of becoming indispensable would have been done already.

The distance from this seems infinite and the relevance to Joao Dellacruz in his Brazilian electronics workshop in Sao Paulo problematic[3]. He feels lonely and dispensable indeed, his situation being much worse than Lyell's. For eight years now, he has been working on the design of a new electronic MOS chip, profiting from a joint venture of industry, the military government and the university, all of which wanted Brazil to be self-sufficient in building computers. Joao and his boss argued at the time that it was also necessary for Brazil to become independent in the manufacture of chips, and that it was better to start with the most advanced designs so as to leapfrog the older generations of chips. They were given a small amount of money to equip a workshop, and to explore the architecture of other MOS chips devised in American and Japanese universities.

For a year or two they thought they would be at the centre of a huge nationalist movement for creating a 100 per cent Brazilian computer. Their workshop would become the obligatory passage point for technicians, students, the military, electronicians from industry. 'He who controls the chips', they used to quip, 'will rule the computer industry.' Unfortunately they were the only ones convinced of this order of priority. The military wavered and no limitation was imposed on the import of foreign chips – only on the import of computers. Joao's lab was no longer the centre of a possible industrial venture. The imported chips were cheaper and better than any of those they could design. Moreover, they were bought and sold by the thousand while Joao and his boss, now deprived of a possible alliance with industry, could devise only a few prototypes and had no customers to help with debugging.

The two electronic engineers then tried to become the centre not of an industry but of some university research. Joao switched his goals and decided to work on a PhD. The problem was that there were no other professors working on MOS chips in Brazil. Luckily, he then got a fellowship to go to Belgium where his boss had studied. Joao worked hard on a very small stipend, so small that, after two years, he had to return to Sao Paulo. Once back there, matters got really bad. The instruments with which he had studied his chips in Louvain were so much better than the ones he had in his workshop that none of the results he had obtained in Belgium were reproducable in Sao Paulo. The intricate circuitry was simply invisible. To make matters worse, he soon learned that his boss – who was also his thesis supervisor – was so disgusted by the state of Brazilian research that he had decided to leave for a position in Belgium. Five years after the beginning of his study, Joao had not one page of his thesis written. His only treasure consisted of a few precious wafers made according to the MOS process. 'With this,' he thought, 'I will always be able to start a small industry if my luck turns.' In the meantime the Japanese were now selling MOS chips which were a hundred times more powerful than his. Furthermore, the state committee had rejected his grant application for a new automated chip designer, arguing that there were not enough researchers in the field to justify the expense. The reader will have an idea of Joao's state of despair if they know that the inflation rate was now 300 per cent while his already small salary was adjusted only once a semester! Joao was becoming so poor that he was contemplating a third part-time job – in addition to his research and his many private teaching lessons. He was now so rarely in his workshop that his equipment – obsolete anyway – was used by the nearby university for teaching purposes. Still, he was proud of having been chosen by the government to advise them which Japanese firm should be preferred for setting up an automated MOS chip factory somewhere in the north of Brazil . . . .

This is indeed a sad story but certainly more frequent than the success stories told in the earlier chapters. Joao cannot create a speciality, no matter how far outside he goes. His workshop is not at the centre of anything, it becomes the annex of a teaching institution. His thesis is not the text that every other researcher has to quote and to take into account; it is not even written. His chips are not the only design that can hold together the assembled interests of industry, government, the military, consumers and journalists; it has become an obsolete piece of technology, a meaningless prototype no one will put to use. Instead of being able to establish itself as a lab which has become the obligatory passage point for countless other people, Joao's workshop is a place no one needs to pass through. It is not strategically placed between anyone's goal and the fulfilment of this goal, and this means, as we saw in the last chapter, that Joao *interests no one*.

Talking with Joao reveals a yet sadder story. All the people I have presented so far had to resist dissenters. To do so they had to write more technical articles, to build bigger laboratories, or to align many helping hands. But who are the people who Joao may challenge or those who may contest Joao's demonstrations? The government? The military? The state grant committee? No, because all these people take no notice of Joao's work and are all situated *outside* the intricate design of MOS chip circuitry. Could it be his colleagues? No, because he has no colleagues, and those who exist, far away in Japan and North America, are too far ahead to be interested in Joao's work. The only one who could remain interested, his thesis supervisor, has now gone, leaving Joao as the only one in the country with his speciality.

What happens to the inside of a speciality made up of only one person? This is the question that makes Joao so despondent: the inside disappears as well. Since he has no one to discuss the draft of his articles with, no one to try out the links he makes between various parts of chip architecture, no one to whom he can submit his proposals for trials of strength, no one to debug his prototypes, Joao ends up not *knowing* what is real and what is fictional in MOS technology. Using the terms I defined in chapter 2 Joao does not know what is objective or subjective. As with Robinson Crusoe on his island, the boundaries between daydream and perceptions becomes fuzzy, since he has no one to dissent with him and thus create a difference between facts and artefacts. Joao feels that the rhetoric of science I showed in Part I of this book is going the other way round: his papers become less and less technical – he now writes only for news magazines, his arguments become cheaper and cheaper – he avoids discussions with other foreign experts. Joao feels he is out of the proof race and becoming more so every day. To start new research is almost impossible now. His equipment is too old, the Japanese too advanced, and his own knowledge too untried. The speciality, made up of one member, will soon have nothing special in it. Joao will be a 'former engineer' barely surviving by giving lessons and writing popular science articles. He really fears that the speciality will soon have – in Brazil at least – no outside support and no inside existence either.

The first lesson to be drawn from this unfortunate example is that there is a direct relationship between the size of the outside recruitment of resources and the amount of work that can be done on the inside. The less people are interested in Joao's workshop the less Joao knows and learns. Thus, instead of trying out new objects which are then able to hold together the interested groups, Joao shrinks away and comes out of his lab empty-handed.

The second lesson from this example is that an *isolated* specialist is a contradiction in terms. Either you are isolated and very quickly stop being a specialist, or you remain a specialist but this means you are not isolated. Others, who are *as specialised as you,* are trying out your material so fiercely that they may push the proof race to a point where all your resources are barely enough to win the encounter. A specialist is a counter-specialist in the same way as a technical article is a counter-article (Chapter 1) or a laboratory is a counter-laboratory (Chapter 2). It is when the amount of resources is large enough that many counter-specialists may be recruited and set against one another. This dissent in turn elevates the cost of the proof race, multiplies the trials of strength, redesigns new objects which, in turn, may be used to translate more outside interests, and so on. But as long as research in internal combustion engine, neuroendocrinology, geology or chip design does not yet exist as a job, there is no specialist inside and no interested groups outside.

(2) Making the laboratories indispensable

Now that we start realising what happens to science in the making when preliminary groundwork is not made, let us look in the log book of a dedicated layperson who decided to shadow the head of laboratory – henceforth named 'the boss' – situated in California.[4]

March 13: everything is all right, the boss can easily be located at his bench performing experiments on pandorin.

March 14: the boss has spent most of his time in his office answering phone calls from twelve successive colleagues to whom he wrote about his new pandorin (four in San Francisco, two in Scotland, five in France, one in Switzerland) – I could not hear what he said.

March 15: I almost missed the plane. The boss flew to Aberdeen in order to meet a colleague who denies that pandorin is a real, independent substance of any physiological significance. While in Aberdeen, he kept calling all over Europe.

March 16: morning: new plane to the South of France; the boss is welcomed by the heads of a big pharmaceutical concern; I barely got a taxi; they discussed all day how to patent, produce and start clinical trials of pandorin and a host of other substances.

– evening: we stop in Paris to discuss with the Ministry of Health the setting up of a new lab in France to promote research in brain peptides; the boss complains about French science policies and red tape; he writes a list of names of people who could possibly be attracted to this new lab; they discuss space, salaries and work permits; the Ministry promises to relax the regulations for this project.

March 17: the boss has breakfast with a scientist who flew from Stockholm to show him how his new instrument was able to locate traces of pandorin in the brains of rats; the pictures are beautiful; the boss speaks of buying the instrument; the other man says it is still a prototype; they both make plans to interest an industry in manufacturing it; the boss promises to advertise the instrument; he hands out a few samples of pandorin to the other scientist for further testing.

– afternoon: exhausted, I miss the ceremony at the Sorbonne where the boss gets an honorary degree from the university. I arrived in time for the press conference he gives afterwards; the journalists are very surprised because the boss lambasts French science policy; he asks everone to be prepared for a new revolution in brain research, the first harbinger of which is pandorin; he attacks journalists who give a negative image of science and are always after sensations and revolutionary discoveries; over a drink afterwards he proposes to a few colleagues the setting up of a scientific committee that would force journalists to behave and not to freely propagate wild claims.

– night: we reached Washington; I am pleased to see that the boss seems tired too.

March 18: morning: a big meeting in the Oval Office with the President and representatives of diabetic patients; the boss gives a very moving speech explaining that research is soon going to break through, that it is always slow, that red tape is a major problem, and that much more money is needed to train young researchers; parents of diabetics answer and urge the President to give priority to this research and to facilitate as much as possible the testing of new drugs from the boss's laboratory; the President promises he will do his best.

– lunch: the boss has a working lunch at the National Academy of Science; he tries to convince his colleagues to create a new sub-section, he explains that without this all his colleagues in this new discipline are lost either in physiology or in neurology and their contribution is not rewarded as it should be; 'we should have more visibility', he says; they discuss how to vote down another colleague, but I am three tables away and do not hear who he is.

– afternoon: a bit late at the board meeting of the journal *Endocrinology*; I cannot sneak into the room; I just learned from the secretary that the boss complains about the discipline being ill-represented and about bad referees who turn down hosts of good papers because they know nothing about the new discipline; 'more brain scientists should be brought in.'

– on the plane: the boss corrects an article a Jesuit friend asked him to write on the relations between brain science and mysticism; the boss explains that pandorin is probably what gave Saint John of the Cross his 'kick'; he also adds in passing that psychoanalysis is dead.

– late afternoon: we arrive at the university just in time for his course; he ends it by reflecting on the new discoveries and how important it is that bright young men enter this booming field full of opportunities; after the course he has a brief working meeting with his assistants and they discuss a new curriculum to include more molecular biology, less mathematics, more computer science, 'it is crucial,' he says, 'that we get people with the right training; the ones we've got now are useless.'

– evening: (blank, too exhausted to follow)

March 19: when I arrive the boss is there already! I had forgotten that it was the day of the site visit for one of his grants, a one-million-dollar affair; the visitors are having discussions with everyone, probing every project; the boss remains aloof in his office 'so as not to influence either the visitors or the staff'. I miss the official dinner.

March 20: morning: the boss is in a psychiatric hospital trying to convince doctors to set up a first clinical trial of pandorin on schizophrenics; unfortunately the patients are all so loaded up with drugs that it will be hard to isolate the effect of pandorin; he suggests that the doctors and himself write a co-authored paper.

– afternoon: we roam around a slaughter house; the boss tries to convince the head of the 'hatchet crew' – I don't know the technical term – to try another way to hack sheep heads off so as not to damage the hypothalami; the discussion seems hard; I am so nauseated that I don't hear a word.

– late afternoon: the boss gives a good dressing down to a young postdoc who did not draft the expected paper on pandorin in his absence; he decides with his collaborators which of the next generation of high pressure liquid chromatography to buy; he goes on perusing the new figures obtained this afternoon on a more purified sample of pandorin.'

We may stop the reading of the log book at this point. Even if it is a busy week it is far from an unusual one. Following a scientist may turn out to be a tiring job and may force the follower to *visit* many parts of the world and many more groups in our societies than expected: high officials, corporations, universities, journalists, religious figures, colleagues and so on.

How could we define the boss's way of doing research from 13 to 18 March? To answer this question we should consider another dedicated layperson who, during the same week, shadowed not the boss but one of his collaborators. Contrary to the first inspector, this one did not move from the laboratory; she stayed all week, twelve hours a day at the bench or in the office submitting pandorin to the sort of trials we described in Chapter 2. If she answered a few phone calls they were from the boss or from colleagues engaged in the same task in other institutions, or from suppliers. Asked about her boss's trip she seemed a bit condescending. She wants to stay at arm's length from lawyers, industry or even government. 'I am just doing science,' she says. 'Basic science, hard science.'

While she stays in the laboratory the boss moves around the world. Is the boss simply tired of bench work? Or is he too old to do worthwhile research – this is often what is muttered in the coffee breaks inside the laboratory? The same grumbles greet West's constant politicking in Kidder's story.[5] West is always moving around from headquarters to marketing firms and from there to electronic fairs. While he is away, the microkids are working like devils, completely insulated from any economic or political hurdle. Each of them works just on one microcode.

This case shows how important it is to decide who are the people to study. Depending on which scientist is followed, completely different pictures of technoscience will emerge. Simply shadowing West or the boss will offer a businessman's view of science (mixture of politics, negotiation of contracts, public relations); shadowing the microkids or the collaborator will provide the classic view of hard-working white-coated scientists wrapped up in their experiments. In the first case we would be constantly moving *outside* the laboratory; in the second, we would stay still deep *inside* the laboratory. Who is really doing research? Where is the research really done?

A first answer comes when the two observers sent to study the boss's lab put together their log books at the end of a year-long observation. They note that the collaborator got a paper accepted in a new section of the journal *Endocrinology* – a section created by the boss; that she has been able to employ a new technician thanks to a special fellowship from the Diabetic Association – after the speech given by the boss at the White House; that she now gets fresh hypothalami from the slaughter house which are much cleaner than before – an outcome of the boss's complaints; that she has two graduate students attracted to her work after they had taken the boss's course at the university; that she is now contemplating a position offered to her by the French Ministry of Health to set up a new laboratory in France – thanks to long negotiations of the boss with French high officials; that she has got a brand new instrument from a Swedish firm to map minute amounts of peptides in the brain – in part because of the boss's involvement in setting up the company.

To sum up, she is able to be deeply involved in her bench work *because* the boss is constantly outside bringing in new resources and supports. The more she wants to do 'just science', the costlier and the longer are her experiments, the more the boss has to wheel around the world explaining to everyone that the most important thing on earth is her work. The same division of labour happens with West and his team. It is *because* West has been able to convince the Company to let them try the *Eagle* project that the young men are able to devise, for the first time in their careers, a brand new computer. The more they want to work 'just on technical matters', the more people West has to seduce.

The consequence of this double move is a trade-off between the intensity of the drive to interest people 'outside' and the intensity of the work to be done 'inside'. As we saw in the last chapter, this trade-off is due to the fact that the interest of all the 'interested' people will last only if, for instance, the new computer and the new pandorin may tie all of them together and become the obligatory passage point for pursuing their usual work. To do so, *Eagle has to be* fully debugged and pandorin *has to be* an undisputable fact; when West's overselling and the boss's bluff are called, all the data they showed should withstand the trials of strength. Because of this trade-off between what has been promised outside and what holds inside, an enormous pressure is then diverted back to the collaborators. They all have to work hard and to submit *Eagle* and the pandorin to all possible trials; to buy the best equipment, to recruit the best graduates. It is whilst submitted to this enormous pressure that they say 'we are just doing science'.

The first lesson to be drawn from these examples looks rather innocuous: technoscience has an inside because it has an outside. There is a positive feedback loop in this innocuous definition: the bigger, the harder, the purer science is inside, *the further outside other scientists have to go.* It is because of this feedback that, if you get inside a laboratory, you see no public relations, no politics, no ethical problems, no class struggle, no lawyers; you see science isolated from society. But this isolation exists only in so far as other scientists are constantly busy recruiting investors, interesting and convincing people. The pure scientists are like helpless nestlings while the adults are busy building the nest and feeding them. It is because West or the boss are so active outside that the microkids or the collaborator are so much entrenched inside pure science. If we separate this inside and this outside aspect, our travel through technoscience would become entirely impossible. At each crossroads, we would not know whom to follow. On the contrary, it is clear that we have to do like Kidder and, from now on, split our attention and follow both the purely technical – as we did in Chapters 1, 2 and 3 – and, so to speak, the 'impurely' technical. Our old friend the dissenter of

Chapters 1 and 2, or the fact-builder, were so stubborn only because other people outside were busy at work; we have yet to follow these people.

(3) What is technoscience made of?

I have portrayed three very contrasting situations: in the case above the science to be studied was clearly divided into a vast inside part – the laboratories – and a large outside part orchestrating the recruitment drive; in the first two cases scientists were struggling to *create a difference* between an inside speciality – in which they could then work – and an outside mixture of contradictory interests – that cut through their speciality and threatened to destroy it entirely. However different the three examples, two featurtes remained constant. First, the ability to work in a laboratory with dedicated colleagues depended on how successful other scientists were at collecting resources. Second, this success in turn depended on how many people were already convinced by scientists that the detour through the lab was necessary for furthering *their own goals*.

(A) 'WHO IS REALLY DOING SCIENCE, AFTER ALL?'

What do the words 'their goals' mean? As we know, they designate an *ambiguous* translation of scientists' and other people's interests. For instance, if the boss is so successful when talking to the Ministry, the President, the Diabetic Association, his students, his lawyers, the head of a pharmaceutical industry, newsmen and fellow academicians, this means that *they* think they are furthering *their* goals when helping him to extend *his* lab. The same thing is certainly true with West. His group is enthusiastic about building a new computer and beating the North Carolina research centre; for this they are all ready to work twelve hours a day seven days a week. Still, at the end, it is Data General's share of the market that is increased and it is De Castro, the big boss, who is more pleased than any other. The young kids' interests, those of West, of De Castro and of the Data General Board of Directors were all aligned, at least for a few months. This *alignment* is precisely what is lacking in the two other examples. The Church, the universities, the gentry, the state, the public, the amateurs , the fellow geologists, all have mixed feelings about letting Lyell develop an independent geology; when Lyell talks about his interests, no one else at first feels that he means 'their interests' as well. Difficult negotiations are still going on to keep all these contradictory wills in line. In Joao's case, it is clear that the interests are all at loggerheads. When he talks about his goals, no one else in the whole world thinks they are theirs as well: neither the military, nor industy, nor his colleagues. The relation between Joao and the others is so *unambiguous* that no community of interest is possible.

So, to sum up, when scientists and engineers are successful in creating a vast inside world, it means that *others* are working towards more or less the same goal; when they are unsuccessful, it means that scientists and engineers are left *alone* to pursue their direction. This sounds like a paradox: when scientists appear to be fully independent, surrounded only by colleagues obsessively thinking about their science, it means that they are fully dependent, aligned with the interest of many more people; conversely, when they are really independent they do not get the resources with which to equip a laboratory, to earn a living or to recruit another colleague who could understand what they are doing. This paradox is simply the consequence of the feedback mechanism I presented in the two sections above: the more *esoteric* a piece of technoscience the more *exoteric* has to be the recruitment of people. This sounds like a paradox only because we sever the two aspects; so, we tend to think that a poorly funded workshop is more tied to outside interests than a well-funded one, whereas it is poor because it is less tied; conversely, when we visit a gigantic cyclotron we tend to think that it is more remote from anyone's direct interest, whereas it is remote only because of its tight links with millions of people. This mistake occurs because we forget to follow simultaneously the inside and outside scientists; we forget the many negotiations that the latter had to carry over for the former to exist at all.

Let us ponder a minute on this inverse relationship. Are we not running into a major difficulty, which could stall our journey through technoscience, if we ask *who is really doing science?* If we say 'the people who work in the labs of course', we know from the example of Lyell or of Joao that this answer is grossly incomplete since by themselves they could not even earn a living or set up a controversy. So we have to *complete* the list of people who are doing science. But if we include in the list all the supporters necessary to transform isolated and helpless scientists into people like West or the boss, we run into an apparent absurdity: shall we say that De Castro, the Ministry of Health, the Board of Directors, the President, are all *doing science?* Certainly yes, since it is to convince them that West and the boss worked so hard for their lab; certainly not, since none of these convinced supporters works at the bench. So we are in a quandary over what seems two equally ridiculous answers. Since our goal is to follow those who are doing technoscience, our enquiry is checked if we can no longer decide who is really doing the work!

Of course, if we follow through the logic of the first answer we can get out of the difficulty. This method, which is accepted by most analysts, is precisely the one we cannot use. It involves saying that the long list of people who support the laboratories constitute a necessary *precondition* for technoscience to exist as a pocket of pure knowledge. In others words, although all these people are necessary to provide resources, they are not shaping the very content of the science made. According to this view there is a real boundary to be drawn between the inside and the outside. If you follow the outsiders you will meet a series of politicians, businessmen, teachers, lawyers and so on. If you stay inside you will get only the nitty-gritty of science. According to this division, the first crowd has to be taken as a sort of necessary evil for the second to work quietly.

The consequence is that whatever knowledge you may gain about the one crowd can teach you *nothing* about the other: the cast of characters and the plots they are enmeshed in will be totally different. This divorce between context and content is often called the internal/external division. Scientists are inside, oblivious to the outside world that can only influence their conditions of work and its rate of development.

I hope it is clear to readers that if they were to accept this division, it would be the end of our trip. All our examples have sketched a constant shuffling to and fro between outside world and laboratory; now an impassable barrier is thrown up between the two. I have implicitly suggested, and will now give the skeleton of, a different anatomy of technoscience: one in which the internal/external division becomes the provisional outcome of an inverse relationship between the 'outside' recruitment of interests – the sociogram – and the 'inside' recruitment of new allies – the technogram. With each step along the path the constitution of what is 'inside' and what is 'outside' alters.

There are two solutions to the problem of the grossly incomplete definition of science against the incredibly broad: either throwing up a theoretical and impassable barrier between 'inside' and 'outside', or tracing an empirical and variable limit between them. The first solution gives *two different stories* depending on where you start – and brings this book to a close; the second solution provides *the same story* in the end no matter if you start from the outside or the inside – and allows this book to go on!

(B) EVERYBODY IS MADE TO GIVE A HAND

To decide between the two versions, let us go back to the second section and trace a simplified map of the boss's travels. Remember that 'doing science' meant two different things for the collaborator working inside the lab and for the boss travelling outside. However, it was clear from the example that they were both doing science since the resources diverted by the boss were then activated by the collaborator; conversely, each new object squeezed from the lab by the collaborator was immediately converted into resources by the boss, so as to secure newer and fresher sources of support. This process, pursued by the collaborator and the boss at the same time, has the shape of a loop or of a cycle. However, as we saw in the first section, this loop may turn *inward* or *outward*: the science may shrink so much that there is no distinction between collaborator and boss, and soon after no new object and no supporter; or it may turn in the direction that makes the science *grow*. What does this mean? As I show in Figure 4.1, it means that more and more elements are part of the cycle. I have artificially divided these elements into money, workforce, instrument, new objects, arguments and innovations, and sketched only three complete cycles.

Let us start with the people who provide money. At the beginning the boss is simply receiving funds; in the middle circle he is heading many national committees that decide who should receive the money; at the end, he is part of the state establishment that legislates on how much money should be given, to which science, and through which system the funds should be allocated and controlled. At the beginning few people have their fate linked to the boss's enterprise; at the end, quite a lot.

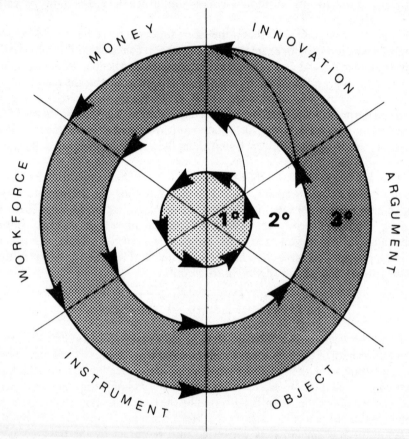

Figure 4.1

Going counter-clockwise, we meet the workforce the boss needs to recruit once he has the money. At the beginning he does the whole job with his own hands and eyes; in the middle he recruits young kids already trained by the university departments or the technical schools; at the end, he is heading new departments, new universities, and advocating major changes in training and priorities throughout the educational system. He may even go further, writing textbooks, giving public lectures, enthusing his audience with a zeal for his science. From the start to the end, the boss has had to go further and further afield, recruiting more and more people and attaching his enterprise to that of more and more schools.

Still further around the circle, we meet the instruments which are so important for shaping new objects. When the whole process is very small-scale, the boss is using only the instruments available or ones he can tinker with; in the middle, he may be devising new instruments and advising industry on prototypes; at the end, he is on the Board of several companies that build instruments, advocating their use in hospitals, fighting the legislation that limits their spread; or in the case of other sciences we may find him at Hearings urging Congress to help in the planning of gigantic new instruments. Here again, we started with few people interested in the boss's cycle, and we may end up with a whole branch of industry tied to his fate.

Further round the circle we encounter the trials produced by the collaborators using the instruments. At first, very few allies are brought in; in the middle, more unexpected ones are put into the picture; at the end, inside huge laboratories, undergoing terrible and unexpected trials, new objects are shaped by the thousand. As we saw in Chapter 2, the more the laboratory grows the wider is the mobilisation process of non-human elements for which the scientists speak.

Next, we encounter the arguments. As we have already learned in Chapters 2 and 3, the boss at first utters weak non-technical claims only which are difficult to publish at all; in the middle, his increasingly technical articles are accepted faster and faster by many more technical journals of higher status; at the end, the boss creates new journals, advises publishing firms, advocates the creation of new data banks, and exhorts colleagues to set up professional associations, academies or international organisations. What started as a timid and controversial claim ends up as an incontrovertible and well-established body of knowledge or a respectable profession.

We then meet the innovations. At the onset, the boss is barely able to convince anyone to use his arguments, his substances or his prototypes. They stay in his small lab like Joao's chips. In the middle, more and more people have been sufficiently interested by the boss to lend their force to his projects: many hospitals, many other disciplines are putting the arguments to good use, spreading the innovations further. At the end, the boss is on the Board of several companies, heading many committees and is the founder of several associations which are all facilitating the spread of the innovation as much as they can. What was limited to one man's lab now circulates through long networks everywhere in the world.

Finally, we come full circle to the beginning of the diagram. At first, the boss is too weak to obtain more grants, more space and more credit simply on the grounds of his previous activities. In the middle, his work becomes recognised, his articles and those of his collaborators are read and quoted, his patents enforced; grants, space and prestige may be more easily secured. At the end, all the forces enrolled through the process are ready to attribute the responsibility of their general movement to him and to his lab or his discipline. What at first had been an isolated place has become by the end an obligatory passage point. By this time, whatever the others do or want, the boss's lab grows – see translation 5 in Chapter 3.

No matter how simplified this general picture is, one thing in it is clear: growth comes from *tying together* more and more elements coming from less and less expected sources. At some point in section 2, we saw slaughter houses, the French Ministry of Health, the Oval Office and brain peptides having a bearing on each other. It is utterly impossible to delineate an outside border to the picture – in which only 'context' for science would be encountered – and an inside core – in which only 'technical content' would be produced. It is easy, on the contrary, to see how the laboratory has to become more and more technical in order to attach so many and so disparate elements to one another. What is clearly separated in the first version – that is, the internal and the external – is precisely what has to be attached so tightly in the second.

If we agree to the superiority of the second version over the first, then another lesson may be drawn from this example. When I write that many people, institutions, instruments, industries and new objects are tied to the boss's enterprise, this means two things at once: first that they are tied to the boss whose lab has become an obligatory passage point for them, but also that *he is tied to them*. He had to go far out of his way to fetch them; he had to bend over backwards to recruit them. If not he would not have risen at all. Thus, when we glance at Figure 4.1 we do not see either the boss's story or the story of the enlisted elements; we see the story of all of them *when they get together and share a common fate*. Those who are really doing science are not all at the bench; on the contrary, there are people at the bench because many more are doing the science elsewhere. The time has now come to turn our attention towards these other people.

Part B
Counting Allies and Resources

In the preceding part we solved two difficulties. First we learned that in our trip through technoscience we should follow simultaneously those who stay inside the labs and those who move outside, no matter how different the two groups appear. Second, we learned that in the construction of technoscience we have to include all the people and all the elements that have been recruited or are doing the recruiting, no matter how foreign and unexpected they seem at first. Is it possible to get an idea of who the people are who are making technoscience and how the various roles are distributed among them?

To answer this question we are going to use the statistics that professional bodies gather in various countries – but especially in the United States – in order to control or to develop what they call Research and Development.[6] No matter how crude and often biased or inaccurate these statistics are, they provide us with at least an order of magnitude. They map out for us the strongholds and the weak points of technoscience. Instead of presenting individual cases as I have done so far, we are now going to get an idea of the *scale* of technoscience simply by using the statistics of the many institutions that manage scientists.

(1) Counting on Scientists and Engineers

The most striking figures come from the most general statistics: those who call themselves scientists and engineers in the census are much fewer than those interested by and enrolling them in the construction of facts and machines. In the United States they are only 3.3 million people (*Science Indicators* 1982, (*SI*) 1983,

Table 4.1

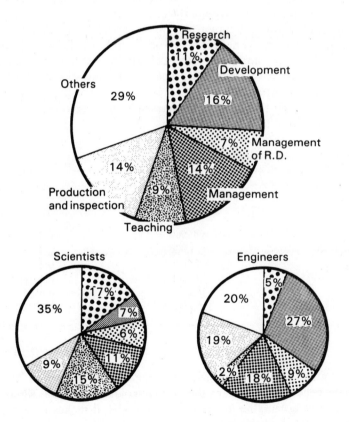

Science Indicators 1982, Figure 3.6, pp. 66.

p. 249), no matter what degree they hold and what work they do. Only 3.3 million say they have some familiarity with any of the black boxes. The 250 million others are supposed to have the barest knowledge provided them by primary or secondary school.

If we wish to consider those who are said to have participated in the definition and negotiation of some black boxes, the number decreases drastically. Most people with a training in science and engineering do not do research or development. In the US, for instance – the country for which we have the most figures – slightly more than a quarter of all scientists and engineers are engaged in R & D.

Table 4.1 is the sort of table that shows the absurdity of the diffusion model criticised at the end of Chapter 3. If we were to believe that bench workers are the only ones 'really doing science' we would have to take into account only some 900,000 people for the US (that is the first two dark areas in the pie charts in Table 4.1); all the others, that is three-quarters of all scientists and engineers, can be forgotten! For the translation model, however, researchers are clearly the tip of the iceberg; many more are needed who work outside in order to make the inside possible, and those who help in the definition, negotiation, management, regulation, inspection, teaching, sale, repair, belief and spread of the facts are part and parcel of 'research'.

The impossibility of limiting technoscience to full-time researchers is clearly demonstrated by Table 4.2:

Table 4.2

Number of scientists and engineers engaged in R & D as a proportion of the workforce		
	Number of scientists and engineers	*Number of scientists and engineers/workforce*
US (1981)	890,000	0.59%
England (1978)	104,000	0.4 %
France (1979)	73,000	0.32%
Germany (1977)	122,000	0.46%
Japan (1981)	363,000	0.65%
USSR (81 estimate)	1200,000	0.90%

Science Indicators 1982, table 1.3 p. 193.

Two and a half million of scientists and engineers cannot make 700 million other people believe and accept all the hard facts of technoscience. Although this disproportion is acceptable in the diffusion model, it makes no sense in the translation model. And this ridiculously small figure has been obtained in the most favourable case. We considered only the most industrialised and richest countries of the North, lumping together *all the disciplines* and introducing no further distinction between Research and Development. Moreover, since the developed countries account for around 90 per cent of all the R & D in the world (94 per cent of the money and 89 per cent of the workforce according to O.E.C.D.[7]), it means that, when travelling through the vast world, one would have one chance in 1500 of meeting someone who has an active role in shaping beliefs and technics. It would mean that only 3 million people are disseminating beliefs and machines, enlisting the 5 billion people on the planet! Quite an extraordinary feat, which means either that these few people are superhuman or that we were wrong in limiting the fact construction to scientists. Many more people than the few scientists officially recognised as such ought to be engaged in shaping technoscience.

It is possible to push the apparent paradox created by the small number of scientists much further. Being counted in the statistics as engaged in R & D does not mean that as many people have had the sort of experience I pictured in Chapter 1 and 2, that is, a direct familiarity with the writing of a technical article, with the setting up of a controversy, with the shaping of new allies, with the devising of new laboratories. If we take the possession of a Phd as an indication of a close and long familiarity with technoscience *in the making*, and if we limit the number of scientists and engineers to the number of doctorates engaged in R & D, the figures we arrive at are much smaller still.[8] If the construction of facts was limited to the research done by doctorates, it would mean that only 120,000 persons in the United States would make the 250 million others believe and behave, enrolling and controlling them in accepting newer and harder facts. One man would be able to enrol and control 2000 others! And, again, this figure has been obtained by lumping all the sciences and all the technics together without any distinction between research and development.

The paradox created by the diffusion model grows to massive proportions if we try to distinguish occupations and disciplines inside the remaining tiny numbers. Remember that in Table 4.1 we saw that only 34 per cent of all scientists and engineers in the US were engaged in R & D or managing it but more than 70 per cent of all the scientists and engineers engaged in R & D are working in industry.[9] So, even the tip of the iceberg is not made of what is commonly called 'science'. If we wished to become closer to the cliché of pure disinterested science we would have to consider only doctorate holders employed by universities or other public institutions and doing research, that is limiting technoscience to academics. If we do so, the figures shrink still further.[10] The number of people who most closely resemble what is commonly called 'scientist' – basic research in a non-profit institution – in the US amounts to something like 50,000 (full-time equivalent). This figure is obtained by rolling all the sciences into one. This is not the tip of an iceberg any more, it is the tip of a needle.

When we talk about 'science' the readers might think of famous scientists in highly prestigious disciplines and universities having produced new revolutionary ideas and products which are now believed, used and bought by hundreds of millions of people. People like Lyell, Diesel, Watson and Crick come to mind. However, considering technoscience as made up of these people is as impossible as making the pyramid of Cheops balance upside down. The great men and women of science to whom prestige accrues are simply too few to account for the gigantic effects they are supposed to produce.

Still, we have chosen the best conditions in order to measure the scale of technoscience. Had we made fewer *ad hoc* assumptions, this scale would be much smaller. For instance, all our figures come *after* a long period of expenential growth in R & D spending and in the training of scientists and engineers.[11] The official size of technoscience would be limited to much smaller numbers had we measured it *before* this boom. No matter how prestigious are the Galileos, the Newtons, the Pasteurs and the Edisons, they were still more isolated and scattered in their own time and societies than the relatively large armies of professional researchers of today. The sciences, which seem so small compared to the number of people they claim to enrol and control, nevertheless dwarf their past so much that they can be said to have almost *no past*. As far as numbers are concerned, technoscience is only a few decades old. The famous scientists studied so much by historians of science can all be found in the minute tail of an exponential curve. To parody Newton, we could say that technoscience is a giant on the shoulders of dwarfs!

There is a second supposition which provides us with an inflated view of technoscience. I made the supposition that all the academic scientists who most resembled the cliché of a scientist were all *equally good*. Even if science was made of a mosaic of small clusters, I assumed that all the clusters were equal. But this is far from being so. There are huge inequalities even inside the small number of academic scientists. There is what is called a *stratification* among scientists.[12] This asymmetry modifies what is called the visibility of a scientist or of a claim.[13] When discussing controversies and dissent, proof race and translations, I have always assumed that each claim and each counter-claim was highly visible and stimulated the debate. This was too favourable a presentation. The vast majority of the claims, of the papers, of the scientists, are simply *invisible*. No one takes them up, no one even dissents. It seems that even the beginning of the process has not been triggered off in most cases.

There is not only a stratification among scientists' productivity, there is also a stratification in the means for making science. From Chapter 2 and from Joao's example, we know that all laboratories are not equal before God. The ability to pursue a dispute depends crucially on the resources one is able to muster on one's side. These resources are concentrated in very few hands. First, this is visible inside the same country.[14] Disputing a fact, launching a controversy, proposing an article *outside* of the top institutions becomes much more difficult, and the more so the further you are away from them. We know why from Chapters 2 and 3: the cost of the proof increases at each turn of the controversy; those who are not able to follow the proof race in their own labs and who still wish to argue have either to break their way into the top institutions or to quit the game altogether.

This stratification is visible inside the same country, but it is still more visible between developed countries.[15] Half of technoscience is an American business. All the other developed countries work on smaller chunks of science. Since hard new facts are made by mustering resources and holding allies in line, the stratification in manpower, money and journals means that some countries will enrol, and others will be enrolled. If a small country wishes to doubt a theory, reject a patent, interrupt the spread of an argument, develop its own laboratories, choose its own topics, decide on which controversy to start, train its own personnel, publish its own journals, search its own data base, speak its own language, it might find this impossible. The same situation I described in Chapter 1 between Mr Anybody and Mr Manybodies may be found between countries with a big share in R & D and countries with a very small share in it. Like Mr Anybody, the country with a small system of science may believe the facts, buy the patents, borrow the expertise, lend its people and resources, but it cannot dispute, dissent or discuss and be taken seriously. As far as the construction of facts is concerned, such a country lacks *autonomy*.[16]

After quickly surveying the figures sketching the scale of technoscience, we clearly understand that limiting it to 'insiders' would lead us to a complete absurdity. We would soon be left with a few hundred productive and visible scientists, in a handful of richly endowed laboratories generating the totality of all the facts believed and of all the machines used by the 5 billion people living on this planet. The distribution of roles made by the diffusion model has been really unequal: to the happy few is reserved the invention, discussion and negotiation of the claims, while the billions of others are left with nothing else to do but to borrow the claims as so many black boxes or to remain crassly ignorant. Scientists and engineers are too few, too scattered, too unequally distributed to enrol and control all the others. Limited to their own force they could not secure the strongholds so necessary to render relevant their rhetoric. For the diffusionists, this conclusion is not a problem, as we saw in Chapter 3: 'on the contrary,' they argue, 'if scientists are so few and do such extraordinary things, it is simply that they are the best and the brightest; these few isolated minds see what Nature is and are believed by all the others because they are right.' Thus, for them, all the figures above do not raise any major problem, they simply *add* to the prestige of a few scientists isolated in the midst of so much obscurity and ignorance!

(2) Not counting only on scientists and engineers

The first section presents us with a picture that may be interpreted in two opposite ways: either the few really good scientists are endowed with the demiurgic powers of making millions believe and behave, or they are scattered in marginal spots, lost in the midst of multitudes who could not care less. However, we know from Part A that this alternative is *also* that of the scientists themselves. West, Diesel, the boss, or Joao, depending on what they do and who they recruit, may be endowed with demiurgic powers – since everyone goes through their labs – or stay marginal figures unable to influence anyone's work. We also learned in the first part that, in order to decide between the demiurgic interpretation and the marginal one, we should not consider only those who call themselves scientists – the tip of the iceberg – but those who, although they stay outside, are nevertheless shaping the science and form the bulk of the iceberg. Now that we have beaten at its own game the diffusion model that asserted that scientists, ideas and prototypes were the only important part of science, we should no longer hesitate to reintroduce all the participants excluded from the official definition of real research into the picture. But how can this be done, since, by definition, statistics on manpower only list those who are officially doing science? There is, fortunately, in the same statistics, a simple way of measuring the multitudes enrolling scientists; they do not appear under the guise of manpower, but under that of *money*. Even distorted in statistics, budgets are a fair estimate of the amount of interest scientists have been able to secure for their work.

If we consider the most aggregated figure available, not on the personnel but on the money, we gain one order of magnitude (Table 4.3).

Table 4.3

Percentage of GNP devoted to R & D	
United States (1981)	2.6%
England (1978)	2.2%
France (1978)	2.6%
Germany (1981)	2%
Japan (1981)	2.4%
USSR (median estimate)	3.6%

Science Indicators 1982 (SI 1983 p. 7)

Table 4.3 gives gross estimates but their general scale is interesting: it means that the few hundreds of thousands of scientists have been able to have a bearing on something like 2.5 per cent of the GNP of the richest industrialised countries.

Does this relatively substantial figure mean that all this money is obtained for the few people that an official rendering of science would consider as 'real scientists'? Not at all, because all kinds of research are lumped together in Table 4.3. The traditional labels to break down statistics are those of basic research, applied research and development. Although it is possible to discuss endlessly the precise boundaries between these terms, we have learned enough in this book to define them for our purpose. As I showed in Chapter 3, obtaining new allies is good, but only insofar as these many allies are able to act as one disciplined whole. Thus, we may distinguish two moments in the recruitment of new allies:

one that multiplies their numbers, and the other that turns them into a single whole. We may call research the first moment and development all the work necessary to make a black box black, that is, to turn it into an automaton that counts as one routine piece of equipment. If we talk of research we will be led more into the sort of situation described in Chapters 1 and 2, with technical papers, discussions, controveries, undisciplined new objects; if we talk of development we will tackle the problems of Chapter 3, putting more emphasis on the hardware and the question of how to discipline the new objects and the people who transfer them. But the distinction is often moot, and should be seen as two aspects of one single strategical problem.

No matter how fuzzy all these distinctions are, the statistics obtained by using them are clear enough, as shown in Table 4.4.

Table 4.4

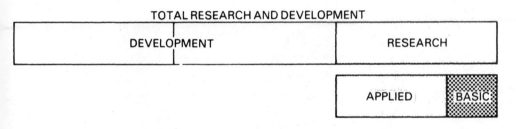

TOTAL RESEARCH AND DEVELOPMENT

Science Indicators 1982 (*SI* 1985, p. 40)

Although the diffusion model would consider only basic science as worthy of attention – the rest flowing effortlessly from it – we see that, by and large, scientists and engineers have been able to gather support only when they do *not* do basic research. Of nine dollars spent, only one goes for what is classically called 'science'. Technoscience is on the whole a matter of development.

Is it possible to go further and to consider who are the supporters of technoscience when it is successful? Remember that, on the one hand, according to our first principle scientists and engineers need many others to build all their black boxes, but that, on the other, they are too few to keep them in line, especially if they wish to make millions of others believe and behave. The only way to solve this problem is for scientists to link their fate to that of other, much more powerful groups that have *already solved the same problem on a larger scale*. That is, groups that have learned how to interest everyone in some issues, to keep them in line, to discipline them, to make them obey; groups for which money is not a problem and that are constantly on the look-out for new unexpected allies that can make a difference in their own struggle. Which groups are these? Another look at statistics gathered in the United States will tell us.

Because these figures are so large-scale they give us an idea of the most important transfers of money, and thus an outline of the main translations of interest (Table 4.5). Essentially, R & D is an industrial affair (three-quarters is carried out inside firms) financed out of tax money (amounting to 47 per cent in the US (*SI* 1983, p. 44)). This is the first massive transfer of interest: scientists have succeeded only insofar as they have coupled their fate with industry, and/or that industry has coupled its fate to the state's. Without this double move technoscience shrinks to minuscule size as we see when only basic science is considered. Now it becomes an affair between the universities and the state: universities do nine-tenths of basic research which is almost totally paid out of the Federal Budget. As can be expected, applied science occupies an intermediary position, 50 per cent being paid by the government and industry and carried out by the universities.

Table 4.5

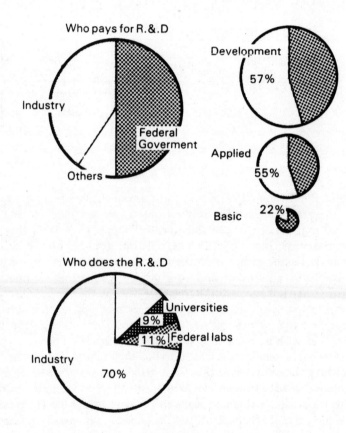

Who pays for R.&.D

Industry

Federal Goverment

Others

Development 57%

Applied 55%

Basic 22%

Who does the R.&.D

Universities 9%

Federal labs 11%

Industry 70%

What sort of topics drain so much taxpayer's money into industry and the universities? The answer is to be found in Table 4.6.

The outsiders are coming into the picture. Defence takes up something like 70 percent of all public R & D spending. Technoscience is a military affair. The only exception is Germany – and Japan, but this exception is itself due to another scientifico-military venture: the dropping of the atomic bombs in 1945, that forced Japan to surrender and to abandon most military research.

Table 4.6(a)

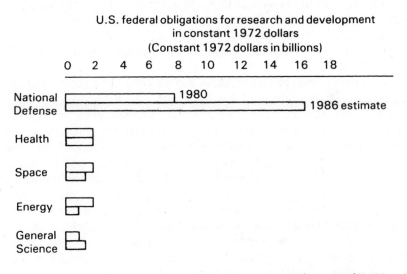

U.S. federal obligations for research and development in constant 1972 dollars
(Constant 1972 dollars in billions)

Science Indicators 1984 (SI 1985, p. 40)

Table 4.6(b) Distribution of government support of R & D by national objective and by country (per cent) in 1980

	US	Japan	W. Germany	France	UK
Defence	63.7	16.8	24.4	49.3	64.8
Health	15.2	11.2	15.3	7.5	3.9
Advancement of knowledge	3.0	4.1	14.2	15.0	12.9
Energy and infrastructure	14.2	34.4	30.9	16.0	10.1
Agriculture	2.7	25.4	2.9	4.3	4.5
Industrial growth	.3	12.2	12.4	7.9	3.8

Science Indicators 1982 (SI 1982, p. 199 and OECD 1982, p. 202)

It is not a strange coincidence or an unwanted evolution that so tightly attaches the development of armies and technoscience. The military obviously foots the bill rather well. For centuries, they have enlisted people and interested them in their action, so much so that most of us are ready to obey them blindly and to give up our lives if required. As far as enrolling, disciplining, drilling and keeping in line are concerned, they have proved their mettle and on a much larger scale than scientists have ever tried.[17] The interested and obedient layperson required by scientists to disseminate their facts is much easier to train than the disciplined soldier ready to sacrifice himself. Besides, the military have been interested in unexpectedly shifting the balance of power with new resources and weapons. It is not surprising then that the few scientists and engineers capable of providing new and unexpected allies capable of changing the balance of power have met with the military frequently during the course of history to promote the production of new weapons.

The similarity between the proof race and the arms race is not a metaphor, it is literally the mutual problem of *winning*. Today no army is able to win without scientists, and only very few scientists and engineers are able to win their arguments without the army. It is only now that the reader can understand why I have been using so many expressions that have military connotations (trials of strength, controversy, struggle, winning and losing, strategy and tactics, balance of power, force, number, ally), expressions which, although constantly used by scientists, are rarely employed by philosophers to describe the peaceful world of pure science. I have used these terms because, by and large, technoscience is part of a war machine and should be studied as such.

This link between war and technoscience should not be limited to the development of weapon systems. To fully grasp it, it is necessary to consider more generally the **mobilisation** of resources, whereby I mean the ability to make a configuration of a maximal number of allies act as a single whole in one place. Research into new weapons is one obvious focus, but so is research into new aircraft and transport, space, electronics, energy and, of course, communications. Most technoscience is concerned with facilitating this mobilisation of resources (see Chapter 6).

The only other big chunk of civilian research visible on Table 4.6 is that of health. Why is it that scientists have been successful in tying their work to this topic? Although it does not fit the bill as well as the army, the health system has done similar groundwork. Like the survival of the body politic, the survival of the body is a subject in which everyone is directly and vitally interested. Since in both cases money is no object, the health budget, like that of defence, is a gigantic treasure chest where spending is made without limit. In both cases interest and spending have been made compulsory by taxes or the social security system, the latter being as big as the state budget in most industrialised countries. The role played by the military in recruiting, drilling, and forcing everyone to be simultaneously interested and obedient has been played for centuries by physicians, surgeons and health workers. Amateurs have been excluded, quacks and charlatans have been forbidden to practise, everyone has been made to take an interest in health problems, legislation has been passed. Most of the work had already been done when life scientists linked their fate to that of health. So it is not surprising that so much research is conducted on the health system. When scientists and engineers are unable to link their work to either of these two budgets, they fare less well. The remainder of all publicly financed R & D is a puny percentage of the total.

The problem of finding resources to pursue the proof race has been historically solved when budding scientists have linked their fate to that of people whose general goal was seen as being approximately the same: mobilising others, keeping them in line, disciplining them, interesting them. If these conditions are not met, groups of scientists may exist, but they will never be able to increase considerably the cost of proof or to multiply the number of their peers. In any event they will never be granted the demiurgic powers of reshaping the world (which, for instance, atomic physicists have). They will be more akin to the older professional role of the scholar. When scientists hold strong positions, many other people are already there who did most of the groundwork.

(3) The fifth rule of method

We started this chapter by asking *who* the scientists and engineers were; we pursued it by *adding* more and more outside people to the making of science; we then stumbled on an inverse relationship that linked the esoteric and the exoteric aspects of science; afterwards, we had to understand that the few people officially called 'academic scientists' were only a tiny group among the armies of people who do science; finally, we came to realise that when the large armies – in the literal sense – that defend either the body politic or the body were not behind them, scientists remained by and large invisible. The drift from the beginning of the chapter to this point is now clear, I hope, to the eyes of the attentive reader: the *enlisting* scientist endowed with the demiurgic power of enrolling and controlling millions of others may now appear as an *enlisted* employee working in industry on military matters. Which one of the two pictures is the more accurate, and which one allows us to learn more about technoscience?

The only possible answer to this question is that neither of the two is correct because the question is not precise enough. Some of the cases we studied have given us the impression that scientists hold enormous powers like West or the boss; other cases suggested the opposite impression like Lyell at the start of his career, or Joao. What did this impression of power or of weakness depend on? On the presence or the absence of *already* aligned interest groups. Although this sounds as paradoxical now as when we first encountered it in Part A, we have to come to grips with it. The few people officially called scientists and engineers seem to carry the day only when most of the groundwork has already been done *by others*. The proof of this is that if the others are not there, or are too far apart, the few scientists and engineers become still fewer, less powerful, less interesting, and less important. So, in all cases, the presence or the absence of many more people than those doing science at the bench should be studied in order to understand *who* those at the bench are and, as we saw in Chapter 3, *what they do*.

How is it that the many others who count so much when providing laboratories with their powers are *discounted* when the time comes to list the personnel of science? They constitute the most important part of technoscience in all the stories I told, so how can they be so easily pushed out of the picture? To answer this we should remember the trials of responsibility defined earlier. To follow these trials a distinction had to be made between the primary mechanism that enlists people, and the secondary mechanism that designates a few elements among the enlisted allies as the cause of the general movement.

The outcome of these trials in responsibility is to allow the picture of technoscience to be completely reversed. Among the million people enlisted by scientists or enlisting them, and among the hundreds of scientists doing applied research and development for defence and industry, only a few hundreds are considered, and to them alone is attributed the power to make all the others believe and behave. Although scientists are successful only when they follow the multitude, the multitude appears successful only when it *follows* this handful of scientists! This is why scientists and engineers may appear alternatively endowed with demiurgic powers – for good or bad – or devoid of any clout.

Now that we can see through this confusion of two different mechanisms, we understand that 'science and technology', from which we started in the introduction, is a figment of our imagination, or, more properly speaking, the *outcome* of attributing the whole responsibility for producing facts to a happy few. The boundaries of science are traced not in terms of the primary mechanism, but only in terms of the secondary one. The recruitment drive remains invisible. Then, when one accepts the notion of 'science and technology', one accepts a package made by a few scientists to settle responsibilities, to exclude the work of the outsiders, and to keep a few leaders. It is fortunate that we decided from the start to study the *activity* of making science and not the definition given by scientists or philosophers of what science consists of. The hard recruitment drives of Diesel, Pasteur, Lyell, of the boss, the many failures of Joao, would have completely escaped our attention. We would have believed in the existence of a science on the one hand, and of a society on the other, which would have rather missed the point! Here again, Janus speaks two opposite languages at once. On the left side he says that scientists are the cause that carried out all the projects of science and technology, while on the right side scientists are striving to position themselves inside projects carried out by many others.

To remind us of this important distinction, I will use the word **technoscience** from now on, to describe all the elements tied to the scientific contents no matter how dirty, unexpected or foreign they seem, and the expression **'science and technology'**, in quotation marks, to designate *what is kept of technoscience* once all the trials of responsibility have been settled. The more 'science and technology' has an esoteric content the further they extend outside. Thus, 'science and technology' is only a sub-set which seems to take precedence only because of an optical illusion. This will constitute our fourth principle.

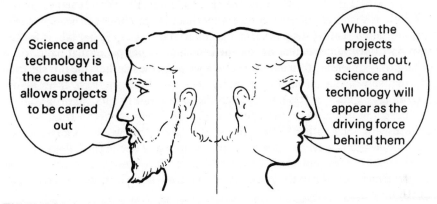

Figure 4.2

There seems, however, to be a danger in extending the size of technoscience, so as to include in it supporters, allies, employers, helping hands, believers, patrons and consumers, because they in turn might be seen as leading the scientists. One might draw the conclusion that if science is not made up of science and led by scientists, it is made up of and led by all the interest groups. This danger is all the greater since this alternative is exactly the one offered by so-called 'social studies of science'. When 'science and technology' is not explained by its internal thrust, it is accounted for by external pushes or demands. Our travel through technoscience should then be full not of microbes, radioactive substances, fuel cells and drugs, but of wicked generals, devious multinationals, eager consumers, exploited women, hungry kids and distorted ideologies. Have we come all this way and escaped the Charybdis of 'science' only to be wrecked on the Scylla of 'society'?

Fortunately, this danger is not a real one if we can see that *all* attribution trials should be cleared away, *including* those which attribute the dynamism of science to social factors. If we are ready to doubt what scientists say about their science, it is not so as to believe what generals, bankers, politicians, newsmen, sociologists, philosophers or managers say about its limit, shape, usefulness or cause of growth. As stated by our fourth rule of method, we should be symmetrical and doubt the boundaries of scientific professions *as much* as those of 'science and technology', no more but no less.

From now on, the name of the game will be to leave the boundaries open and to close them only when the people we follow close them. Thus, we have to be as undecided as possible on which elements will be tied together, on when they will start to have a common fate, on which interest will eventually win over which. In other words, we have to be as *undecided* as the actors we follow. For example, before the boss enters his office, the Minister of Health is still uncertain whether or not it is worth investing in neuroendocrinology; the boss too is uncertain whether or not the Minister will keep the promise made by his counsellors about funding a brand new laboratory; he is also uncertain as to whether or not pandorin is such a revolutionary substance that firm promises can be made to the Minister about curing drug addicts; his collaborator, deep in her lab, is for her own part uncertain whether or not she can claim in her paper that pandorin is biologically different from another substance published earlier; the rats she tried the two substances on might die under the high doses she gave them before providing any answer. It is possible that the collaborator's rats, the drug addicts, the boss, the counsellors, the Minister and the Congressmen will all become aligned with one another so that, in the end, laboratory work has a bearing on national health policy. But it is also possible that any one of these links or all of them might break apart so that the rats die, pandorin becomes an artefact, the Congressmen vote down the budget, the boss irritates the Minister who overrules his counsellors....

The question for us who shadow scientists is not to *decide* which one of these links is 'social' and which one is 'scientific', the question for us, as well as for those we follow, is only this: 'which of these links will hold and which will break apart?' Our **fifth rule of method** will thus be the following: we should be as undecided as the various actors we follow as to what technoscience is made of; to do so, every time an inside/outside division is built, we should follow the two sides simultaneously, making up a list, no matter how long and heterogeneous, of all those who do the work.

After having studied how a weak rhetoric could become stronger, and then how many strong positions had first to be obtained to make this added strength relevant, the time has now come to study those who are not enrolling or are not enrolled by scientists and engineers – that is, all those who do *not* participate in the work of technoscience.

PART III

From Short to Longer Networks

CHAPTER 5

Tribunals of Reason

In the first part of this book we studied how to go from a weak rhetoric to a strong one, and in the second we followed the scientists and engineers in their many strategies as they go from weak points to the occupation of strongholds. If we wanted to summarise the first four chapters, we could say that they showed a fantastic increase in the number of elements tied to the fate of a claim – papers, laboratories, new objects, professions, interest groups, non-human allies – so many, indeed that if one wished to question a fact or to bypass an artefact one might be confronted by so many black boxes that it would become an impossible task: the claim is to be borrowed as a matter of fact, and the machine or the instrument put to use without further ado. Reality, that is what resists all efforts at modification, has been defined, at least for the time being, and the behaviour of some people has been made predictable, in certain ways at least.

Another way of summarising the same four chapters is to show the other side of the coin: such an increase in the number of elements tied to a claim is to be *paid for* and that makes the production of credible facts and efficient artefacts a costly business. This cost is not to be evaluated only in terms of money, but also by the number of people to be enrolled, by the size of the laboratories and of the instruments, by the number of institutions gathering the data, by the time spent to go from 'seminal ideas' to workable products, and by the complication of mechanisms piling black boxes onto one another. This means that shaping reality in this way is not within everybody's reach, as we saw at length in Chapter 4.

Since the proof race is so expensive that only a few people, nations, institutions or professions are able to sustain it, this means that the production of facts and artefacts will not occur everywhere and for free, but will occur only at restricted places at particular times. This leads to a third way of summarising what we have learned in this book so far, a way that fuses together the two first aspects: technoscience is made in relatively new, rare, expensive and fragile places that garner disproportionate amounts of resources; these places may come to occupy strategic positions and be related with one another. Thus, technoscience may be described simultaneously as a demiurgic enterprise that multiplies the number of allies and as a rare and fragile achievement that we hear about only when all the other allies are present. If technoscience may be described as being so powerful and yet so small, so concentrated and so dilute, it means it has the characteristics of a **network**. The word network indicates that resources are concentrated in a few places – the knots and the nodes – which are connected with one another – the links and the mesh: these connections transform the scattered resources into a net that may seem to extend everywhere. Telephone lines, for instance, are minute and fragile, so minute that they are invisible on a map and so fragile that each may be easily cut; nevertheless, the telephone network 'covers' the whole world. The notion of network will help us to reconcile the two contradictory aspects of technoscience and to understand how so few people may seem to cover the world.

The task before us in the last part of this book is to explore all the consequences that this definition of technoscience as a network entails. The first question I will tackle concerns the people who are *not* part of the networks, who fall through the mesh of the net. So far, we have followed scientists and engineers at work; it is necessary for a while to turn our attention towards the multitudes who do not do science in order to evaluate how difficult it is for scientists to enrol them. Given the tiny size of fact production, how the hell does the rest of humanity deal with 'reality'? Since for most of history this peculiar system of convincing did not exist, how did the human race manage for so long without it? Since even in modern industrialised societies the vast majority does not get close to the process of negotiation of facts and artefacts, how do they believe, prove and argue? Since in most enterprises, there has been no scientist or engineer to occupy obligatory passage points, how do ordinary folk go about their daily business *without* science? In short, the question we have to study in this chapter is what is *in between* the mesh of the networks; then, in Chapter 6 we will tackle the question of how the networks are sustained.

Part A
The trials of rationality

(1) Peopling the world with irrational minds

How do the multitudes left out of the networks see the scientists and the engineers, and how do they themselves consider the outside of these networks?

Take for example the case of weather forecasts. Every day, often several times a day, many millions of people talk about the weather, make predictions, cite proverbs, inspect the sky. Among them, a large proportion listen to weather forecasts or glance at satellite maps of their countries on TV and in newspapers; quite often, people make jokes about weathermen who are, they say, 'always wrong'; many others, whose fate has been linked *earlier* to that of meteorologists, anxiously await forecasts before taking decisions about seeding plants, flying planes, fighting battles or going out for picnics. Inside the weather stations, running the huge data banks fed with satellite signals, controlling the reports of the many part-time weathermen scattered over the planet, sending balloons to probe the clouds, submitting computer models of the climate to new trials, a few thousand meteorologists are busy at work defining what the weather is, has been and will be. To the question 'what will the weather be tomorrow?' you get, on one side, billions of scattered commentaries and, on the other, a few claims confronted with one another through the telexes of the international Meteorological Association. Do these two sets of commentaries have a common ground? Not really, because, on the one hand, the few claims of the meteorologists are utterly lost among billions of jokes, proverbs, evaluations, gut feelings and readings of subtle clues; and because, on the other hand, when time comes to define what the weather has been, the billions of other utterances about it count for nothing. Only a few thousand people are able to define *what* the weather *is*; only their opinions literally *count* when the question is to allocate the huge funds necessary to run the networks of computers, instruments, satellites, probes, planes and ships that provide the necessary data.

This situation creates a rather curious balance account: the weather and its evolution is defined by everyone on earth and the few weathermen provide only a few scattered opinions among the multitudes of opinion, taken more seriously in only small sectors of the public – the military, the ship and air companies, agricultural concerns, tourists. However, when you put all these opinions in one balance of the scale and in the other the few claims of the meteorologists, the balance tips on the side of the latter. No matter how many things are said about the weather, no matter how many jokes are made about the weathermen, the weather of the weathermen is strong enough to discount all the other weathers. If you ask the question 'was it a normal summer or an exceptionally hot one?' although everyone says, everyone feels that it has been a hot summer, the lived opinions of the multitude may be discounted *inside* the networks of the International Meteorological Association. 'No,' they say, 'it was a summer only 0.01 degree above average.' The certitudes of billions of people have become *mere opinions* about the weather whose essence is defined by the few thousand meteorologists. 'You *believed* it was a hot summer, but it was *really* an average one.'

The balance of forces may be tipped in one direction or in another depending on whether we are inside or outside the network developed by weathermen. A handful of well-positioned men of science may rout billions of others. This will happen only, however, as long as they stay *inside their own networks*, because, no matter what the meteorologists think and do, every one of us will still think it was a hot summer and make jokes, the morning after, about the weather forecasts which were 'wrong as usual'. This is where the notion of network is useful: meteorology 'covers' the world's weather and still leaves out of its mesh almost every one of us. The problem for the meteorologists will then be to *extend* their networks, to make their predictions indisputable, to render the passage through their weather stations obligatory for everyone who wants to know the weather. If they are successful, they will become the only official mouthpiece of the earth's weather, the only faithful representatives of its vagaries and evolution. No matter how many people are left out, they will never be as credible as the weathermen. How to obtain such a result does not interest us at this point – see next chapter – because what we want to understand is what happens to everyone's opinion about the weather when meteorologists become the only mouthpiece of weather.

All other predictions become, in the eyes of the scientists, illegitimate claims about the weather. Before meteorology became a science, they say, everyone was fumbling in the dark, spreading half-truths about the shape of clouds or the flight of the sparrows, believing in all sorts of absurd myths mixed up, fortunately, with a few very sound practical recipes. A more charitable interpretation is that they could not get the whole picture and reacted only to local and provisional signs. We now get on the one hand **beliefs** about the weather, and, on the other, **knowledge** of this weather. This is the first time in this book we have paid any attention to these words, and it is important to realise why they have arrived so late, and only to characterise how scientists inside a powerful network see the outside of it. In their view beliefs are more subjective, that is they tell as much about who holds them as about the weather itself; knowledge, on the contrary, is objective, or at least tends to be always more so, and tells us about what the weather is, not about who the weathermen are. Even if beliefs happen sometimes to be in accordance with knowledge, this is an accident and does not make them less subjective. In the eyes of the people inside the networks, the only way for someone to know about climates and their evolution is to *learn* what the climatologists have discovered. People who still hold beliefs about the climate are simply unlearned.

In this rendering of the non-scientists' opinions, a subtle but radical transformation occurs. We are no longer faced with our original asymmetry between the inside and the outside of a network, between the access to satellite maps, data banks, meters and probes, and the access to subtle clues in the garden, to folklore and to proverbs. Resources necessary to make credible claims about the weather are slowly pushed out of the picture. Indeed there is still an asymmetry, but it has progressively become of an entirely different nature: it is now an asymmetry between people who hold more or less distorted beliefs about something, and people who *know* the truth of the matter (or will soon know it). A partition is made between those who have access to the nature of the phenomena, and those who, because they have not learned enough, have access only to distorted views of these phenomena.

The question to raise, in the eyes of the scientists, is not the one I started with: how can so few meteorologists extend their networks to control the definition of what the weather is, in spite of the multitude of contradictory definitions? The question to raise now is this one: how is it that there are *still* people who believe all sorts of absurdities about the weather and its evolution when it is so easy to learn from us what the weather really is? What is surprising is no longer how so few well-equipped laboratories may come to discount and displace billions of others, but how people may *believe* things they could *know* instead.

What one should study and what one should marvel at is now dramatically altered. Many of the questions scientists of various disciplines raise when they think about the outside of their networks are now of a different form: how can someone *still* believe this? Or how can someone have taken *so long* to realise this was wrong? For instance, an astronomer will wonder why 'modern educated Americans still believe in flying saucers although they obviously do not exist'. A modern sociobiologist will be 'interested to know why it took so long for biologists to accept Darwin's theory'. A psychologist would wonder 'why there are people who are silly enough to still believe in parapsychology which has been proven wrong for decades'. A geologist will be incensed by the fact that 'in 1985 there are people who still believe more in Noah's Flood than in geology'. An engineer will wish to receive an explanation of 'why African peasants to this day are refusing to use solar-powered water pumps which are so much more efficient and cheap'. A French physics teacher will be baffled by the discovery that 'nine out of ten of his pupils' parents believe the sun revolves around the earth'. In all these examples it is implicitly assumed that people should have gone in one direction, the only reasonable one to take but, unfortunately, they have been led astray by something, and it is this something that needs explanation. The straight line they should have followed is said to be *rational*; the bent one that they have unfortunately been made to take is said to be *irrational*. These two adjectives, which are the staple of discourse about science, have not been used here so far. They appear only when an assumption is made by scientists about why there are non-scientists. This assumption is pictured in Figure 5.1.

Since what surprises scientists is how people are pushed out of the right path they should have taken, they need to explain these distortions by appealing to special forces (vertical arrows in the diagram). People should really have understood straight away what the reality is, had outside events not prevented them from doing so. 'Prejudices', for instance, may be used to explain 'why Americans still believe in flying saucers'. 'Differences in culture' may be used to

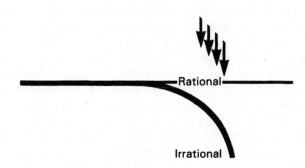

Figure 5.1

account for the fact that 'Africans do not like the use of solar pumps'. 'Outright stupidity' may serve to explain why a colleague behaves so irrationally. Sexual or racial differences may be of some use as well. 'Social explanations' may also be mobilised to account for 'the resistance of biologists to Darwin's theory'. They were probably afraid of the social consequences that such a theory might have in terms of social unrest among nineteenth-century workers. Psychological problems are convenient to use as well because passions may blind people to reason, or unconscious motives may distort even the most honest person. The gamut of explanations that may be provided is extensive and we are not interested in setting up a list which anyway would resemble a gallery of monsters. What interests us in these appeals to outside forces is simply that they come only when one accepts the scientists' position distinguishing between beliefs and knowledge.

By this argument, what is in need of an explanation is only the part of the line that leads away from the straight one. The straight path itself, that is 'rational knowledge', is not in need of any explanation. To be sure, one might find some reasons why weathermen know exactly what the weather is, or why biologists finally learned about evolution, or how geologists discovered continental drift, but none of these explanations bears on the content of the knowledge; they are simply *conditions* leading to, or helping to get at, this content. Since rational knowledge – the straight line – is about what the phenomena are and not about the people who describe them, the only explanations necessary to account for the presence of these claims are *the phenomena themselves* (see Chapter 2, Part C). This happy situation is not the same for irrational claims; they tell us very little about the phenomena and a lot about the people who persist in believing them. Thus, special explanations are required to account for their persistence. This is what David Bloor calls an asymmetric explanation.[1]

A more asymmetric rendering of Figure 5.1 is made by scientists inside their networks. Since the phenomena themselves are the only explanations of rational knowledge, what is needed to discover them? Resources? Allies? Laboratories? Interested groups? No, because these elements that we have studied for five chapters and that make the claim credible have been entirely put out of the picture and no longer have any bearing on the content of science. What is needed to follow the right path is just *a sound mind and a sound method*. What is necessary, on the other hand, to account for the distorted path taken by the believers? Lots of factors which can be chosen from among a long list including 'culture', 'race', 'brain anomalies', 'psychological phenomena' and, of course, 'social factors'. Now the picture of non-scientists drawn by scientists becomes bleak: a few minds discover what reality is, while the vast majority of people have irrational ideas or at least are prisoners of many social, cultural and psychological factors that make them stick obstinately to obsolete prejudices. The only redeeming aspect of this picture is that if it were only possible to *eliminate* all these factors that hold people prisoners of their prejudices, they would all, immediately and at no cost, become as sound-minded as the scientists, grasping the phenomena without further ado. In every one of us there is a scientist who is asleep, and who will not wake up until social and cultural conditions are pushed *aside*.

The picture of technoscience which we have unfolded so far has now entirely disappeared, to be replaced by a world peopled with irrational minds, or with rational minds but victims of more powerful masters. The cost of producing arguments has vanished as has the proof race. Seeing the phenomena face to face does not cost a penny; only good minds free from prejudice are necessary. Nothing makes the extension of knowledge to everyone on earth impossible, it is simply a question of clearing away the distorting beliefs. We may understand why until now I have tried to avoid the notions of belief, knowledge, rationality and irrationality. Whenever they are used they totally subvert the picture of science in action, and replace it by minds, phenomena and distorting factors. If we wish to continue the study of the networks of technoscience, we must straighten up the distorted beliefs and do away with this opposition between rational and irrational ideas.

(2) Reversing the outcome of trials in irrationality

In the last section, I asserted that there was a series of questions we should not try to answer, like 'how come such and such people believe such and such a statement?', since these questions are the consquence of an *asymmetric* treatment by the scientists themselves of what sort of people non-scientists are. To try to answer them has no more meaning than wondering how come a friend of yours did not give your money back, when in fact you did not lend him any money; or explaining how Hermes manages to fly his with small wings before being certain that this god exists and flies! Questions about causes do not deserve an answer if the existence of the effect is not proven first. There would be no special factor to discover for *why* people believe irrational things, if this irrationality was simply a consequence of looking from the inside of the network to its outside – after having bracketed out all the resources necessary for this network to exist, to extend and to be maintained. There is no use in having a discipline like the sociology of knowledge, that tries to account for non-scientific beliefs, if all questions of irrationality are merely *artefacts* produced by the place from which they are raised.

One way to avoid asymmetry is to consider that 'an irrational belief' or 'irrational behaviour' is always the result of an *accusation*. Instead of rushing to find bizarre explanations for still more bizarre beliefs, we are simply going to ask who are the accusers, what are their proofs, who are their witnesses, how is the jury chosen, what sort of evidence is legitimate, and so on, setting up the complete frame of the tribunal in which the accusation of irrationality takes place. Instead of putting the cart before the horse and condemning someone without due trial, we are going to follow the trial for irrationality, and only if a

verdict of guilty becomes unavoidable will we look for special reasons to account
for these beliefs.

The jury (generally small) of this tribunal is made up of the enlightened public
of the Western world. Self-appointed prosecutors parade before this jury, filing
accusations for breaking the laws of rationality (the straight lines of Figure 5.1).
At first, the accusations seem so terrible that the jury is incensed and is ready to
condemn without further ado.

Case 1: There is an hereditary rule in Zande society that states that whenever
someone is a witch he or she transmits this characteristic to his or her offspring.[2]
This should create new rounds of accusations that would run down family trees
and could bring to trial not only the first witch, but also his or her sons,
granddaughters, parents and so on. Not so, the anthropologist Evans-Pritchard
noticed with puzzlement. Instead of drawing this logical conclusion, the
Azande simply consider that there are 'cold' witches in the clan – who are
innocent and not subject to accusation – and that the dangerous 'warm' witch
may be insulated from the rest of the clan. Thus a clear contradiction of the laws
of rationality is presented to the jury. The Azande apply two opposite rules at
once: rule 1: witchcraft runs in the family; rule 2: if one member is accused of
being a witch, this does not mean the rest of his or her clan are witches. Instead of
seeing this contradiction and fighting against it, the Azande simply *do not care.*
This indifference is shocking enough to warrant the accusation of irrationality
made by Evans-Pritchard against the Azande. However, together with the
prosecution, he also enters a plea of attenuating circumstances: if the Azande
were to consider each member of the witch's clan as a witch, the whole clan would
be extirpated, which would threaten the whole society. Thus, *to protect their
society* they prefer not to draw rational inferences. This is illogical, says the
prosecution, but is *understandable:* a social force has taken precedence over
reason. The penalty should not be too harsh, because the Azande are not like us,
they prefer to protect the stability of their society instead of behaving rationally.
As expected from section 1, an explanation has been found as to why some people
have been pushed out of the right path.

Case 2: The prosecution is not so benign with the Trobriad Islanders.[3] Not
only do these tribes have an incredibly complex land tenure system but the
litigation about land that sometimes brings them to court shows constant
breaches of even the most basic principles of logic. Their language is so
inarticulate that it even lacks specific words for linking propositions with one
another. They are unable to say things like ' if . . . and if . . . then . . .'. They do not
understand causality. They do not even have an idea of what is before and what is
after a given proposition. They are not only illogical; not even prelogical, they are
altogether alogical. The court perceives their discussion as a chaotic rambling of
disconnected statements spiced at random with words like 'therefore' 'because'
and 'thus' and mixed with meaningless words in tiresome tirades like this one:

'Therefore I came to reside in Teyava and saw my sister at a different veranda. I had worked hard with them, for our mother. But because my sister had no one, I said to myself, "Oh, this is not good. I will do a bit of kaivatam of course." People of Tukwaukwa I eat your excrement, compared to your gardens the one I made for her was so small. I met her needs, so to speak. I held Wawawa. I held Kapwalelamauna, where today I garden Bodawiya's small yams. I held Bwesakau. I held Kuluboku' (Hutchins 1980, p69.)

Ferreting out attenuating circumstances for the Trobrianders is a hopeless task, and so is the search for the social forces that could explain such a disorderly state of mind. A staunch penalty should be sought for these people who should be cut off from the rest of rational humanity, and imprisoned for life in their islands, unless they entirely recant their errors and start to learn seriously how to think and to behave.

Case 3: The next case is much less dramatic, but still shows a break away from the right path of reason. In the 1870s Elisha Gray was hot on the heels of Alexander Graham Bell in the invention of the telephone, except that Gray was pursuing a multiplex type of telegraph and *not* the telephone.[4] Gray almost discovered the telephone many times over in his career but every time he started drafting a patent for it, his more serious concern for telegraphy led him astray. For him as well as for Bell's father, father-in-law and financial supporters, the telegraph was the technology of the future, whereas the telephone was at worst a 'kid's game' and at best 'a scientific curio'. A few hours after Bell had presented his patent in 1876, Gray deposed a preliminary patent called a 'caveat'. Even at this time, he did not think of seriously fighting in court to contest Bell's priority. Even when Bell offered his patents for sale for $100,000 the managers and advisers of the Western Union – among whom Gray was the most prominent – declared they were not interested. They decided to mount a legal battle against Bell's patent eleven years later when, in 1877, everyone in the Western Union realised, a bit late, that the telephone had a future and that this future would always hamstring Western Union's development. Gray clearly missed the boat and lost his trials in court against Bell's priority, as well as those in history against Bell's wisdom. The prosecution is not without an explanation for this. Gray, they say, was an expert in telegraphy, one of the directors of Western Union and a well-known inventor. Bell, on the other hand, was very much an outsider and he was a complete amateur to the field since his job was to re-educate deaf and dumb people. Bell saw the right path without being blinded by prejudice, whereas Gray, who could have followed the same path and almost did invent the telephone, had been led astray by the weight of his vested interests. The final verdict is not of irrationality, but of lack of openness – outsiders, as is well known, are better than experts at innovation. The penalty, although light at first, is heavy in the long term: everyone remembers Bell's name, but very few have heard of Elisha Gray, who had the 'disadvantage of being an expert'.

Stories such are these are constantly told and retold, passed along, embroidered with many more details, making people laugh or rear up indignantly. Irrationality seems to be everywhere, in savage minds, in children's minds, in the popular beliefs of the lower classes, in the past of scientific or technical disciplines, or in the strange behaviour of colleagues in other disciplines who missed the boat and were led astray. When these stories are told, it really seems that the verdict of irrationality is without room for appeal and that the only question is what penalty should be given, depending on any attenuating circumstances.

It is very easy, however, to *reverse* such an outcome by offering cases for the defence.

Case for the defence 1: There is, in our modern societies, a very strong law that forbids people to kill one another. People who break this law are called 'murderers'. There is also a not-so-infrequent practice that consists of dropping bombs on people who are your enemies. The pilots of these aircraft should therefore be called 'murderers' and brought to trial. Not so, a Zande anthropologist sent to England notices, with some puzzlement. Instead of drawing this logical conclusion, the English simply considers that these pilots are 'murderers in the line of duty' – they are innocent and not brought to trial – and that the other 'wilful murderers' are dangerous and should be tried and imprisoned. Thus, a clear case of irrationality is presented to the same jury who had to decide on the Zande's lack of judgement. From the point of view of the African anthropologist, the English apply two rules at once; rule 1: killing people is murder; rule 2: killing people is not murder. Instead of seeing this contradiction and trying to solve it, the English simply *do not care*. This scandalous indifference offers a clear ground to warrant a trial for irrationality called 'Reason versus the English people'. To be sure, attenuating circumstances may be found for such irrationality. If pilots were brought to court, it would be the destruction of military authority, which would threaten the whole fabric of English society. Thus, to *protect their social institutions*, English people prefer not to draw logical inferences. Here again, social reasons are brought in to explain why such behaviour is not in conformity with the laws of logic.

By proposing a story built with exactly the same structure as that of the prosecution but *symmetric* to it, the defence reverses the clear-cut impression of irrationality. Now, it is the jury who is wondering whether the English are not as irrational as the Azande or, at least, as indifferent to logic because they prefer protecting their cherished social institutions.

Case for the defence 2: The defender, Edwin Hutchins, rises on behalf of the Trobriand Islanders and offers a commentary on the 'rambling tirade' so derided by the prosecution.

Motabesi pleads before the tribunal his right to cultivate a garden he does not own. His sister owns a garden but she has no one to cultivate it. Thus, it is quite responsible for Motabesi to take on her gardening. Does Motabesi really 'eat the excrement' of the Tukwaukwa people? Does he really make such a small garden? No, but it is polite towards the people who are hearing his case to underrate himself and his own garden. This is what is called in court rhetoric 'captatio benevolentiae'. Then Motabesi states his rights to all the gardens he has been given. The garden which is the centre of dispute is called 'Kuluboku'. One of them named 'Kapwaleleamauna' has been given to him by the same lady, Ilawokuva, who owns the garden that is in dispute. This is not a strong inference, and the litigant does not claim it is so, but it is a good point on his behalf. Does Motabesi talk irrationally? No, he simply states a set of connected conditions in support of his case. This is quite reasonable, given the extreme complexity of the land tenure system which is unwritten and has no less than five different degrees of what we Westerners simply call 'ownership'. (Adapted from Hutchins, 1980, p.74).

In the tribunal of rationality, the defender has modified the jury's opinion about the alogical nature of the Trobrianders' minds, by *adding* the *context* of the discussion and the land tenure system on which the reasoning applies to the recorded task. As soon as this is put back into the picture, all the cognitive abilities denied by the prosecution are back also. Trobrianders manage in court like we do, but they have a different land tenure system and they talk in a language unfamiliar to us. It is as simple as that. Nothing very extraordinary, and certainly there are no grounds here for accusing anyone of irrationality, and still less for condemning or proposing penalties.

Case for the defence 3: The story of Bell, the outsider amateur who outstrips Gray, the established expert, is moving and touching, but has been interrupted too soon, the defender says. Were we to continue the story, a completely different outcome would be revealed. We would have hardly heard of Bell if, in 1881, the nascent Bell Company had not bought the Western Electric company and made it its exclusive manufacturer for all its telephone hardware, thus making a standardisation of the telephone network possible at last. But who was the founder of the Western Electric Company? Gray himself, who made numerous other inventions of telephone and electric equipment. Moreover Bell, the imaginative outsider, soon had to leave his own company to be replaced by a great many *specialised experts* in electricity, physics, mathematics, management and banking. If not, the Bell Company would have disappeared in the jungle of the more than 6000 telephone companies that were mixing up their cables and lines all over America at the turn of the century. The amateur triumphed once, but lost out. Thus, if one wishes to explain why Gray missed the telephone and Bell got it in 1876, it is fair enough to explain also why Bell missed the development of his own company ten years later and was pushed aside gently but firmly by experts. The same blindness to the logic of the phone system and its spread may not be used for why Bell won and also for why he eventually lost. It is certainly impossible to use the 'well-known superiority of outsiders in innovation' since this factor would have to be used positively in 1876 and negatively ten years later, the same cause explaining simultaneously the acceleration and the deceleration of the Bell Company! It is equally impossible to explain, by the same attachment to tradition and vested interest, why Gray missed the telephone and why he succeeded in making the Western Electric Company so instrumental in the development of the telephone. Here again the same 'cause' would have to be used alternatively to explain the resistance to innovation and its acceleration. . . .

The jury has reversed its verdict against Gray, simply because the defender let the story go on a bit longer, showing how each factor used to 'explain' a distortion from the right path of reason was later used to 'explain' its opposite. This suggests that it is the whole business of finding 'causes for distortions' that is fatally flawed.

Instead of looking for explanations as to why people hold strange beliefs, the first thing to do, when told one of these many stories about someone else's irrationality, it to *try to reverse their outcome*. This is always feasible by at least one of these means:

(1) Tell another story built around the same structure, but one that applies instead to *the society of the story teller* (shifting for instance from the English anthropologist in Africa to the African anthropologist in England).

(2) Retell the same story but invoke the *context* every time there seems to be a hole in the reasoning and show what sort of unfamiliar topics the reasoning applies to (add for instance to the Trobrianders' rhetoric their complex land tenure system).

(3) Retell the same story but frame it differently by letting it go on longer. This reframing usually renders most of the 'explanations' unusable because, given the right time scale, these explanations are offered for contrary examples as well.

(4)Tell another story in which the rules of logic are broken as well, but this story is not about beliefs but about knowledge held by the story teller. The audience then realises that their judgement was not based on the breaking of the rules, but on the *strangeness* of the beliefs.

When any of these tricks is employed, or combined together, the accusation of irrationality is reversed. There seems to be no case in which an articulate lawyer cannot convince the jury that the others are not so much illogical as simply *distant* from us.

(3) Straightening up distorted beliefs

The task of the jury which has to hear the trials for irrationality becomes rather difficult. At first sight, each case was clear-cut since there seemed to be no difficulty in tracing a *divide* between belief on the right hand and knowledge on the left; no difficulty either in placing derogatory adjectives on the right-hand side – such as 'irrational, gullible, prejudiced, absurd, distorted, blinded, closed', etc. – and laudatory ones on the other – like 'rational, sceptical, principled, credible, straightforward, logical, open-minded', and so on. At the end of the first round of pleading, there seemed to be no problem in defining science by one set of these adjectives and non-science by another set. By adding to the adjectives adverbs like 'purely', 'completely', 'strictly', 'utterly', 'totally', the divide is stressed still more. Once the defendants have talked back, however, the clear-cut divide becomes increasingly fuzzy. Each of the adjectives from one side jumps back to the other side of the divide.

Take the adjective 'sceptic'. At first it nicely defines, for example, Jean Bodin's careful plea for applying good methodology in science and legal matters.[5] If you let the story go on, however, you read that Bodin's sceptism is applied to those who *doubt* witchcraft, so that, in the end, free enquiry in science is for Bodin a way of definitively *proving* the existence of witchcraft against sceptics. Descartes, on the other hand, one of the founders of scientific method, is clearly against all beliefs that cannot resist what he calls 'methodic doubt', belief in witchcraft being one of them. However, even Descartes does not stay very long on the right side of the divide, because he obstinately fills space up with the vortices and denies any form of action at a distance (like gravitation), this running directly contrary to Newton, whose empty space and unmediated gravitation he regards with the same horror as belief in witchcraft and 'occult qualities'. So maybe, after all, we have to consider that Newton and only Newton is on the right side of the divide, all the others before him having lived in the darkness of non-science. But this is impossible as well, because Newton is derided by continental scientists as a reactionary who wants to put mysterious attraction back into the picture and who lacks the most basic principles of scientific method, that is a sceptical and unprejudiced mind. Besides, Newton believes in alchemy at the very time when he writes the *Principia Mathematica*.[6] The only way to stop adjectives jumping randomly from one side of the divide to the other would be to believe that only *this year's* scientists are right, sceptical, logical, etc., thus asking the jury to believe those who plead *last*. But this would be quite an illogical belief since, next year, new scientists will have come along who, again, will have to reprimand their predecessors for having been unfaithful to the rules of scientific method! The only logical conclusion of such an illogical belief being that eventually no one on earth is durably rational.

The jury is by now in a state of despair. If you get clever enough lawyers there is no absurd episode in the history of religion, science, technology or politics that cannot sound as logical and understandable as any other on the good side of the divide, and, conversely, no sound one that cannot be made to look as bizarre as the worst episode on the bad side of the divide. Besides the four rhetorical tricks seen in section 2, it may simply be a question of choosing the right adverbs and adjectives. Bodin, for instance, is considered as an obscurantist, who fanatically believed in witchcraft out of pure prejudice: the proof that old women were indeed witches was for him that they admitted to be so and confessed in writing their flying to Sabbaths; such 'proof' was obtained under torture and in contradiction to the most basic scientific principles since it meant these old women's bodies were simultaneously lying on their litter and dancing with the devil; a simple look at these women asleep would have convinced Bodin of the absurdity of his prejudices. Galileo, on the other hand, courageously rejecting the shackles of authority, arrived at his mathematical law of falling bodies on purely scientific grounds, putting aside the so-called 'proofs' of Aristotelian physicists, and deduced by theory what his experiments showed imperfectly, thus reversing everything the Church believed about the make-up of the universe. Clearly, Bodin is to be placed on the darker side of the divide and Galileo on the more enlightened one. But what happens if we reverse the adverbs and the adjectives? Bodin, for instance, becomes a courageous champion of the faith, who deduced the existence of witchcraft on purely theoretical grounds; he carefully extrapolated from the various experiences undergone by witches' bodies under torture, resisting their many devilish tricks to avoid confession, and discovering a new scientific principle according to which bodies may simultaneously fly and be at rest on their litters. On the other hand, Galileo Galilei, a fanatic fellow traveller of protestants, deduces from abstract mathematics an utterly unscientific law of falling bodies that entails the absurd consequence that all bodies, whatever their nature, fall towards the ground at the same speed; a simple look at daily experience would have convinced Galileo of the absurdity of his prejudice, but he held to it obstinately and blindly against the age-old authority of common sense, experience, science and Church teachings! Who is now on the dark side of the divide, and who is on the enlightened one? Which one of the readers, sitting on the Roman Inquisition, would have let Galileo go and would have put Bodin under house arrest?

There are only two ways to get out of this situation. One is to use derogatory and laudatory adjectives and their accompanying adverbs whenever it suits you. 'Strictly logical', 'totally absurd', 'purely rational', 'completely inefficient', thus become *compliments or curses*. They do not say anything any more on the nature of the claims being so cursed or complimented. They simply help people to further their arguments as swear words help workmen to push a heavy load, or as war cries help karate fighters intimidate their opponents. This is the way in which most people employ these notions. The second way is to recognise that these adjectives are so unreliable that they make *no difference* to the nature of the claim, each side of the divide being as rational and as irrational as the other.

How can we do away with a distinction which is so clear-cut and so fuzzy at the same time, between rational and irrational minds? Simply by retracing our steps. Remember that it was only in the first section of this chapter that we invented the notion of irrational minds, by treating differently what was inside the scientific network and what was outside. This invention was depicted in Figure 5.1 by first supposing a straight line; then, *by comparison* with this line, we noticed a bend out of the right path of reason; finally, in order to explain this bend that in our opinion should not have occurred, we looked for special factors and, in consequence, were dragged to this tribunal of reason where we got embroiled in the sophistry of lawyers. This succession of events all depended on one original move only: *the tracing of a straight line* in Figure 5.1. If we erase it, the whole confusing and unrewarding debate around rationality and irrationality is phased out.

Let us go back to the first case and its rebuttal. The English anthropologist argued that the Azande were faced with a contradiction and avoided it in order to protect the peace of the society. To this, the Zande anthropologist replied that the English were also ignoring contradictions when maintaining simultaneously that the killing of people is murder and that pilots who drop bombs are not murderers. In Figure 5.2 I have drawn the two cases on each side of a dividing line. The two straight dotted lines are traced by the two anthropologists who both offer *ad hoc* social factors to explain the distorted beliefs of the other society. A plane of symmetry divides the picture. According to this image each of the two cases is as illogical as the other[7].

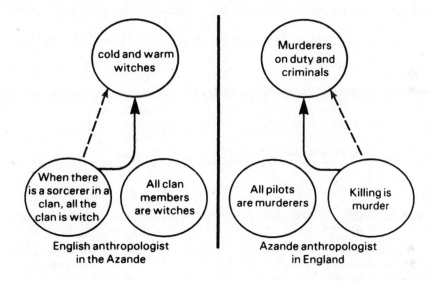

Figure 5.2

There is, however, a major blunder committed by the Zande anthropologist about our Western culture. He made the supposition that when we apply the rule 'killing people is murder' we *implicitly* include the situation of war in the notion of 'killing people'. Then, when we refused to say so *explicitly*, the anthropologist triumphantly argued that we were unable to reason logically. But this is not so, because our notion of murder *never implied* the situation of war – except in very rare cases like those of the Nuremberg trials, which showed how difficult it is to try soldiers who 'just obeyed orders'. So, we cannot be accused of refusing to draw a logical conclusion, if the premisses of the reasoning are in the anthropologist's head and not in ours. It is not our fault if the anthropologist does not understand the meaning of the word 'murder', and is not familiar with its definition in the West. What is wrong in the right side of Figure 5.2 is not our 'distorted belief', it is, on the contrary, the dotted line traced by the Zande anthropologist.

If we feel this to be true for us, we are bound to suspect that it is the same on the other side of the plane of symmetry. The chances are that the Azande never included the possibility of contaminating the entire clan in the definition of the transmission of witchcraft. Here also, the fault is not with the Azande who failed to understand logic, but with Evans-Pritchard who failed to understand the definition of Zande witchcraft.[8] The accusation launched by both the anthropologists about the other's cultures rebounds on them: each is unfamiliar with the culture they study. A breach in logic that whole societies are accused of having committed has been replaced by a lack of familiarity on the part of a few isolated anthropologists sent to a foreign land. After all, this is much more reasonable. It is less surprising to suppose that ignorance made two anthropologists distort others' beliefs than to suppose a whole society bereft of reason.

What shape will Figure 5.2 take if we rub out the anthropologists' mistake?

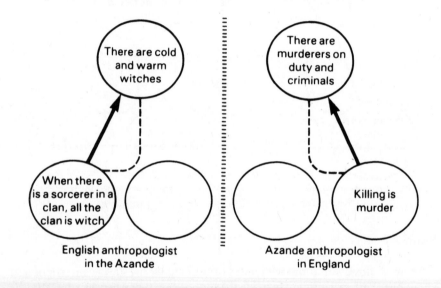

Figure 5.3

The distorted beliefs have now been straightened up. The straight dotted lines invented by the anthropologists, out of ignorance, have been crossed out and so has the 'lack of logic', the 'accusation of irrationality' and the 'social factors' that explained the curves. When the Azande define the contamination of the witch they define 'cold' and 'warm' witches. When we define murder, we distinguish killing 'on duty' and killing 'on purpose'. Period. No one is illogical in this; the definition of a word or of a practice has been traced, that is all. Not the slightest grounds for an accusation of irrationality has been provided here.

The same straightening up may be made for each of the cases we have heard so far. What difference is there between the Trobrianders' logic and ours (see the second case and its rebuttal)? None, but their legal system is different and their land tenure system is foreign to us. What difference is there between Gray's and Bell's mind (see the third case)? None, but they are not interested in the same artifact; one furthers the telegraph, the other the telephone. What difference may be found between the methodologies of Bodin and Galileo? Probably none, each involving elaborate 'laboratory conditions', but they do not believe the same thing and do not tie the same elements together. Cognitive abilities, methods, adjectives and adverbs do not make a difference among beliefs and knowledge, because everyone on earth is as logical or as illogical as any one else. The tribunal declares itself to be unqualified to try the case and decides to free all the people it has arrested. Exit the judges, the jury, the witnesses, and the police. Everyone is innocent of the crime of irrationality, or, more exactly, no one can be *proven* guilty of such a crime. After having peopled the world with irrational minds because we naively wondered why there were so many people who were not scientists, we now understand that it was our wondering that created the problem. Instead of living in a world made of straight dotted lines that people rarely follow, and of distorted paths they most often take because they are carried away by prejudices and passions, we live in a *logical enough world*. People mind their business and get along. . . .

Part B
Sociologics

The lawyers who happen to be articulate, courageous and clever enough to convince the tribunal (a) that all cases of patent irrationality have a lot of attenuating circumstances, (b) that most cases of rational behaviour manifest signs of patent irrationality, and (c) that the tribunal is unqualified and should be disbanded because there is no formal code of laws that could provide a basis for their verdict, are called relativists as opposed to the prosecutors who are called rationalists.[9] Every time a charge of irrationality is filed, relativists argue that it is only an appearance that depends on the jury's relative *point of view* – hence their name – and they offer a new perspective from which the reasoning appears straightforward. Their position is called symmetric and is clearly different from the asymmetric principle of explanation – see above – that looked for social factors to 'explain' the wandering from the right path. Relativists help us understand what falls through the meshes of the scientific network and allow us to resume our journey without being dragged into trials for irrationality.

(1) Running against other people's claims

The problem with these relativists is that, if they were right, we would have to stop travelling here, basking in the contemplation of everyone's innocence. Actually, we could even throw this whole book on a bonfire together with the heavy files of the tribunal proceedings. Why? Because for four chapters we have followed scientists at work who strive to make their claims *more credible* than those of others. So if this enormous work makes no difference they have wasted their time, I have wasted my time, the readers have wasted their time. In taking the asymmetric stand, it is true that the tiny size of scientific networks was ignored, since science and technology were supposed to extend everywhere at no cost, leaving aside only shocking pockets of irrationality that had to be mopped up with better education and sounder methodology. But in the symmetric stand it is the very existence of the scientific network, of its resources, of its ability to sometimes tip the balance of forces, that is utterly ignored. It is not because meteorologists unfairly accuse billions of people of clinging to distorted beliefs about the weather (see Part A, section 1) that we have to deny that, when the time comes to tell the weather, only a few thousand people *succeed* in displacing billions of opinions. The symmetric stand may be more sympathetic and appear fairer, but for us it is as dangerous as the asymmetric one of Part A, since in both cases the very nature of technoscience is negated. It is made either too big or too small, too successful or too unsuccessful.

To be sure, it is the professional duty of lawyers to believe in the innocence of their clients and to convince the jury of it, but lawyers make up only a small part of the system of justice. We should not believe the relativists any more than we believe that no crime has been committed because some good lawyer has obtained the release of his or her clients. Anyway, all lawyers, all relativists, all scientists and engineers, are fighting endlessly to *create* an asymmetry between claims, an asymmetry no one can reverse easily. This is the basis of the lawyer's rhetoric. We have learned from Part A, thanks to the relativists' plea, that this asymmetry should not be accounted for by putting belief (or irrationality) on one side and knowledge (or rationality) on the other. But, still, the problem of accounting for the asymmetry remains intact. If it is no longer the presence or the absence of formal rules of logic that makes the difference, then what is it? Denying that differences are created would be as meaningless as saying 'I will never say no'.

To sum up, the positive aspect of relativism is that, as far as *forms* are concerned, no asymmetry between people's reasoning can be recognised. Their dismissal of the charges always has the same pattern: 'just because you do not share the beliefs of someone you should not make the *additional supposition* that he or she is more gullible than you.' Still, what has to be explained is why we do not all share the same beliefs. The accusation has shifted from form to content.

In a well-known study of unschooled farmers carried out in the Soviet Union, Luria tested their ability to grasp simple syllogisms like this:[10]

'In the far north all bears are white; Novaya Zemyla is in the far north. What colour are the bears there?'

'I don't know. You should ask the people who have been there and seen them,' was a typical answer.

If we were still in Part A, we would see this as a clear failure to grasp the logical nature of the task. This farmer is unable to abstract and to draw consequences from premises (which in logic is called the *modus ponens*). However, the study was replicated by Cole and Scribner in Liberia and they reversed Luria's verdict by employing two of the tactics I have presented at the end of Part A: they let the story go on longer and added the context. Immediately, farmers who had failed similar tests explained their reasoning, by arguing, for instance, that to know the colour of something they would have to see it, and that to see something they would have to be there with the animal. Since they were not there and could not see the animal, they could not tell the answer. This chain of reasoning involves what, in logic, is call a *modus tollendo tollens* (reasoning from the consequent) that is supposed to be more difficult to handle than the other one (reasoning from the antecedent)! There are still differences between what was expected from the test and what the farmers did, but they are not to be found in the *form* of the logic used. Cole and Scribner argue that these farmers have not been to school and that this indeed makes a big difference because most schooling is based on the ability to answer questions unrelated to any context outside the school room. 'Not thinking about the same things' is not equivalent to 'not being logical'. In this example the differences to be looked for have shifted from the form of the claims – 'ability to draw syllogisms' – to their contents – 'number of years at school'. The farmers cannot be accused of being illogical – they use the highly complex *modus tollendo tollens* – but they can be accused of not using logic learned in school; in short they can be accused of not having gone to school. You cannot accuse me of being illogical, but you can belong to another group and wish me out of your way.

From questions about 'minds' and 'forms' we have now moved to questions of clashes between people living in different worlds. One feature of all the episodes I have studied is put back into light: all the accusations were triggered off every time the paths of the accusers and those of the accused *intersected*. Now we can see how to leave relativists to their professional duties as defence lawyers, and continue on our way to simultaneously understanding what the scientific networks capture in their meshes and what escapes them. The weather forecasts over a whole region entail continuous clashes with local people who want predictions about local weather. Hence the reciprocal accusations between meteorologists and local people (Part A, section 1). The two anthropologists – see the first case and its rebuttal – were traversing foreign cultures and were addressing their travel diaries to their colleagues at home in order to settle important debates about rationality. The Trobriander litigants were engaged in a struggle to recover ownership of their gardens; their debates were taped and studied by Hutchins, a Californian cognitive anthropologist, who wanted to go back home with a PhD thesis that would change the opinions which anthropologists have about savage minds – see the second case and its rebuttal. Gray and Bell were extending different networks that were in competition with each other, and their story was told by historians of technology who were not interested in extending the telegraph or the telephone, but who wished to refute arguments about how innovations are favoured or forestalled by social factors (case 3).

As I stressed in Part A, none of these episodes could demonstrate anything irrefutable about the rationality or the irrationality of the human mind. However, they all show that there are many disputes about the weather forecasts, the ownership of gardens, the success of prophecies, the nature of logic, the superiority of telegraph over telephone. These disputes occur inside scientific professions (meteorologists, anthropologists, historians, sociologists); they occur outside of them (about gardens, storms, etc.); they occur at the intersections of the two sets (anthropologists and 'savages', peasants and meteorologists, engineers and historians of technology, etc.). The examples also show that sometimes some of these disputes are settled for a long while: Motabasi got his garden back, Evans-Pritchard's definition of Azande witchcraft remained unchallenged for decades, Hutchins got his PhD, Bell became the eponym of Ma Bell. . . . We have now shifted from debates about reason to disputes about what the world of different people is made of; how they can achieve their goals; what stands in their way; which resources may be brought in to clear their way. In effect, we are back to the beginning of Chapter 1: what can be tied to a claim to make it stronger? How can the claims that contradict it be untied? No one is accusing anyone else of irrationality, but we are still struggling to live in different worlds.

(2) What is tied to what?

We cannot say anything about reason or logic, but whenever we run against other people's claims, we realise that other things are tied to them and we put these links to the test. Let us take three canonic examples of conflicts over *classification* when people try to answer differently the question of what element pertains to which set.

Classification 1: A mother is walking in the countryside with her daughter. The little girl calls 'flifli' anything that darts away very rapidly and disappears from view. A pigeon is thus a 'flifli' but so is a hare fleeing in panic, or even her ball when someone kicks it hard without her seeing it. Looking down in a pond the little girl notices a gudgeon that is swimming away and she says 'flifli'. 'No' the mother says 'that is not a "flifli", that is a fish; *there* is a "flifli" over there', and she points to a sparrow taking off. Mother and daughter are at the intersection of two chains of associations: one that ties a ball, a hare , a pigeon, a gudgeon to the word 'flifli'; the other one that distinguishes a verb 'flee' that could indeed apply to several instances above – but not to the ball – and a noun 'bird' that would apply only to the pigeon and the sparrow. The mother, not being a relativist, does not hesitate to name 'incorrect' her daughter's usage of the word 'flifli'. 'It is one or the other,' she says, 'either a verb or a noun.' 'Flifli' recalls a set of instances that are not usually associated in the mother's language. The girl has to reshuffle the instances gathered so far under the word 'flifli', under the new headings 'bird', 'fish', 'ball' and 'to flee'.

Classification 2: The Karam of New Guinea call 'kobtiy' an animal which is neither a 'yakt', a 'kayn', a 'kaj', nor any of the other names they have for animals.[11] This animal, all by itself in this category, is a strange beast. It lives wild in the forest, it is a biped, it has fur but it lays eggs, it has a heavy skull. When hunted, its blood should not be shed. It is the sister and the cross-cousin of the Karam who hunt it. What is it? This enumeration sounds like a riddle to the anthropologist Ralph Bulmer, who intersects with the Karam culture for a while. He himself calls this animal a 'cassowary' and since it lays eggs, is a biped and possesses wings, Bulmer places it among the birds although it has no feathers, does not fly and is very large. In a typically asymmetric fashion, Bulmer looks for explanations as to why the Karam put cassowaries apart from birds, when they really *are* birds. Once we erase this unfair accusation, however, what we see here are two taxonomies in conflict: one made by the Karam, the other made by the New Zealander; one that is called *ethno*taxonomy or ethnozoology because it is peculiar to the Karam, the other is called simply taxonomy or zoology that is peculiar to all the naturalists *inside* the networks in which their collections are gathered and named.[12] Bulmer has never hunted the cassowary, nor is he running the risk of mating with his cross-cousin – as long at least as he stays in New Guinea. This is not the case with the Karam. They are very interested in this big game and very concerned with incest. Thus, Bulmer sticks to his taxonomy (the cassowary is a bird) and to his research programme (explaining to colleagues why for the Karam a cassowary is not a bird); the Karam also stick to their taxonomy (a kobity cannot be a yakt, that's all) and their hunting and marriage habits (the wilderness is dangerous, so is incest). Associations made between instances of birds are as solid as the two worlds to which Bulmer and the Karam are tied: the Anthropology Association, the journal *Man* and Auckland University in New Zealand on the one hand; the upper Kaironk Valley in the Schrader Mountains of New Guinea on the other.

Classification 3: Ostrom, a well-known paleontologist, wonders whether *Archæopteryx*, one of the most famous fossils, is or is not a bird.[13] To be sure, it had feathers, but did it fly? The problem with evolving from reptiles to birds is in the long intermediary stage where the animal needs to develop feathers, wings, flight muscles and sternum, whereas none of these features is useful *before* it flies – this is called preadaptation. What could be the use of wings and feathers for an animal like *Archæopteryx* that was, according to paleontologists, utterly unable to fly or even to flap and which, if it had glided, would have crashed after a few metres? Ostrom has an answer but it is a quite radical one because it means a reshuffling of large parts of fossil taxonomy and a rethinking of the physiology of the famous dinosaurs. If you take feathers off *Archæopteryx* it looks every bit like a small dinosaur and not at all like a bird. But still it has feathers. What for? Ostrom's answer is that it is to protect this tiny animal from losing too much heat. But dinosaurs are cold-blooded so a thick protection would kill them because they could not take up heat fast enough from the outside! Not so, says Ostrom, dinosaurs are *warm-blooded* creatures and *Archæopteryx* is the best proof of that. Feathers are not there for flying but for protecting a warm-blooded dinosaur from heat loss, allowing it to remain very tiny. Since *Archæopteryx* is not a bird, but is a tiny feathered dinosaur that is only preadapted to flight, this proves that dinosaurs are warm-blooded. It is no longer necessary to search for bird ancestry among the *Pterodactyls* or the crocodiles. It is among the dinosaurs that birds should be placed! Two other paleontologists, in a letter to *Nature*, even suggested doing away with the class of bird altogether. There are now mammals and dinosaurs, of which latter class the birds are living representatives! The sparrow is a flying dinosaur, not a bird; *Archæopteryx* is a terrestrial dinosaur, not a bird. In the midst of the controversy between paleontologists over dinosaurian physiology, the fossil feathers are made to occupy a crucial position. They may allow the champions of cold-blooded dinosaurs to push *Archæopteryx* into the trees and into the class of birds, or the champions of warm-blooded dinosaurs to do away with the birds and to keep *Archæopteryx* on the ground.

In the examples above, each conflict about what is associated to what *traces* what the world of the other people is made of. We do not have on the one hand 'knowledge' and on the other 'society'. We have many trials of strength through which are revealed which link is solid and which one is weak.

The child in the first story above does not know in advance how strongly her mother clings to the definition of 'bird' and of 'to flee'. She tries to create a category that mixes everything that darts away, and she fails every time, confronted by her mother who breaks down this category. The little girl is learning what a part of her mother's world is made of; sparrows, balls and gudgeons cannot all be 'flifli'; this cannot be negotiated. The choice for the daughter is then to give up her category or to live in a world made of at least one element different from that of her mother. Holding to 'flifli' does not lead to the same life as holding to 'birds' and 'to flee'. The girl thus learns part of the language *structure* by trying out what her mother holds to. More exactly, what we call 'structure' is the shape that is slowly traced by the girl's trials: this point is negotiable, this is not, this is tied to this other one, and so on. One sure element of this structure is that 'flifli' has not got a chance of surviving if the girl is to live with English-speaking people.

Bulmer, in the second story, is doing exactly the same thing as the little girl. He is learning both the Karam's language and society by testing the strength of the associations that make it impossible for the Karam to take the cassowary for a bird. Do they mind if Bulmer says it is a bird? Yes, they seem to mind a lot. They throw up their hands in disgust. They say it is absurd. If Bulmer insists, many arguments are brought in as to why it cannot be a bird; the cassowary cannot be hunted with arrows, it is a cross-cousin, it lives in the wilderness . . .. The more Bulmer probes, the more elements are brought in by his informants that prevent the kobtiy from being a yakt. At the end Bulmer realises that the choice for him is either to abandon his association of cassowary with birds or to stay for ever outside the Karam's society. In practice, what he learned through these trials is part of the shape of the Karam's *culture*. More exactly, what we call 'culture' is the set of elements that appear to be tied together when, and only when, we try to deny a claim or to shake an association. Bulmer did not know in advance how strong the reasons were that made kobtiy stand apart from all the birds – especially because other New Guinea tribes were putting it in the category of birds like all Western taxonomists. But he slowly learned that so much was attached to this animal by the Karam that they could not change their taxonomy without a major upheaval of their ways of life.

When Ostrom, in the third story, purports to weaken the linkages between *Archæopteryx* and the living birds he does not know in advance how many elements his opponents are going to bring in to rescue this most famous of evolutionary lines from being broken off. The more he tries to show that it is in fact a warm-blooded dinosaur with a protective coat, the more absurd it seems to the others. A major upheaval of paleontology, of taxonomy, of the organisation of the profession, would be necessary for his argument to be accepted. Ostrom is then confronted with a choice: either to give up his argument or *not to belong to the profession* of paleontologist any more – a third possibility is to redefine what it is to be a paleontologist so that his argument will be part of it. In practice, Ostrom's trials trace the limit of a paradigm, that is the set of elements that have to be modified for some association to be broken away or for some new one to be established. Ostrom does not know in advance what shape the paradigm has. But he is learning it through probing what holds tightly and what gives way easily, what is negotiable and what is not.

What are often called 'structure of language', 'taxonomy', 'culture', 'paradigm' or 'society' can all be used to define one another: these are some of the words used to summarise the set of elements that appear to be tied to a claim that is in dispute. These terms always have a very vague definition because it is only *when* there is a dispute, *as long as it lasts*, and *depending on* the strength exerted by dissenters that words such as 'culture', 'paradigm' or 'society' may receive a precise meaning. Neither the little girl, nor Bulmer, nor Ostrom would have revealed part of the systems of associations of the others had they not dissented or come from the outside, and been confronted by a choice about which group to belong to or which world to live in. In other words, no one lives in a 'culture', shares a 'paradigm', or belongs to a 'society' *before* he or she clashes with others. The emergence of these words is one consequence of building longer networks and of crossing other people's path.

If we are no longer interested in adding to the many little clashes between beliefs, in establishing any grandiose dichotomy – child versus adult, primitive versus civilised, prescientific versus scientific, old theory versus revolutionary theory – then what is left to us in order to account for the many little differences between chains of associations? Only this: the number of points linked, the strength and length of the linkage, the nature of the obstacles. Each of these chains is *logical*, that is, it goes from one point to the other, but some chains do not associate as many elements or do not lead to the same displacements. In effect, we have moved from questions about *logic* (is it a straight or a distorted path?) to **sociologics** (is it a weaker or a stronger association?).

(3) Mapping the associations

We have seen how to be free from the belief in the irrationality of certain claims (Part A), and also from the symmetric belief that all claims are equally credible (sections 1 and 2). We can go on following people striving to make their claims *more* credible than others. While doing so *they map for us and for themselves the chains of associations that make up their sociologics.* The main characteristic of these chains is to be unpredictable – for the observer – because they are totally heterogeneous – according to the observer's own classification. Bulmer pursues what he thinks is a purely taxonomic question and he is dragged into an obscure story about cross-cousins. Ostrom tackles what is for him simply a question of paleontology, and he is led into a huge paradigm shift that renders his reinterpretation of *Archæopteryx* difficult. How are we to study these unpredictable and heterogeneous associations that are revealed by the growing intensity of the controversies? Certainly not by dividing them into 'knowledge' and 'context', or by classifying them into 'primitive' or 'modern' ones, or by ranking them from the 'more reasonable' ones to the 'most absurd'. All actions like 'dividing', 'classifying' or 'ranking' do not do justice to the unpredictable and heterogeneous nature of the associations. The only thing we can do is to follow whatever is tied to the claims. To simplify, we can study:

    (a) how causes and effects are attributed,
    (b) what points are linked to which other,
    (c) what size and strength these links have,
    (d) who the most legitimate spokespersons are,
    (e) and how all these elements are modified during the controversy.

I call sociologics the answer to these questions. Let me take three new examples of what I will call 'free association' – free, that is, from the observer's point of view.

Free association 1: on Christmas Eve of 1976 in the Bay of St Brieuc in Brittany, deep down in the water thousands of scallops were brutally dredged by fishermen who could not resist the temptation of sacking the reserve oceanographers had put aside.[14] French gastronomes are fond of scallops, especially at Christmas. Fishermen like scallops too, especially coralled ones, that allow them to earn a living similar to that of a university professor (six months' work and good pay). Starfish like scallops with equal greed, which is not to the liking of the others. Three little scientists sent to the St Brieuc Bay to create some knowledge about scallops love scallops, do not like starfish and have mixed feelings about fishermen. Threatened by their institution, their oceanographer colleagues who think they are silly and the fishermen who see them as a threat, the three little scientists are slowly pushed out of the Bay and sent back to their offices in Brest. Whom they should ally themselves with to resist being rendered useless? Ridiculed by scientists, in competition with starfish, standing between greedy consumers and new fishermen arriving constantly for dwindling stocks, knowing nothing of the animal they started to catch only recently, the fishermen are slowly put out of business. To whom should they turn to resist? Threatened by starfish and fishermen, ignored for years by oceanographers who do not even know if they are able to move or not, the animal is slowly disappearing from the Bay. Whom should the scallops' larvae tie themselves to so as to resist their enemies?

Answer to these three questions: the Japanese scientists. Yes, it is in Japan that the three scientists saw with their own eyes scallop larvae tie themselves to collectors and grow by the thousands in a semi-protected shelter. So here they are bringing the idea of a collector back with them and trying it in the St Brieuc Bay. But will the Breton larvae be interested by the collectors as much as the Japanese ones? Are they of the same species? Frail links, indeed, those that tie the fate of science, fisheries, scallops, starfish and Japan to that of the St Brieuc Bay. Besides, collectors are expensive, so colleagues and high officials should be convinced in order to give money for new collectors made of all sorts of materials that will eventually be to the liking of the larvae. But when the scientists have convinced the administration, and when larvae start to thrive on their collectors, fishermen cannot resist the temptation of a miraculous catch and they fish the scientists' scallops! So new meetings have to be organised, new negotiations to be started not with the larvae this time, but with the fishermen. Who speaks in their name? They have a few representatives, but without much power. The very spokesmen who agreed to let the scientists work were the first to dredge the reserve on Christmas Eve 1976!

Free association 2: in June 1974 several of us were at a party given in honour of the doctoral thesis of Marc Augé, a French anthropologist, by his main informant, Boniface, on the Alladian littoral of the Ivory Coast.[15] We ate and drank under straw huts looking at the ocean, without swimming in it because Boniface had warned us that the undertow was too dangerous. One of our friends, slightly drunk, went to swim in spite of the warning. Soon he was dragged away by the surf. All of us, blacks and whites, looked helplessly at him. Boniface, an old man feeling responsible for his guests, went to the sea with other, younger friends. Many minutes later the surf brought our friend back to the beach, but for long hours we watched Boniface's body bobbing in the waves. All the village assembled, his family clan, screaming and yelling but powerless. I felt responsible as a white and hated my friend, this other white man, who had caused the tragic drowning of our host. I also feared that the villagers, sharing the same collective interpretation, would turn against us and mob us. I clung protectively to my little daughter. No one looked at us or threatened us in any way, however. The elders of the village wanted simply to know *who* had caused Boniface's death and started a very careful enquiry. At no time did they even think of us. The responsibility had to be somewhere in Boniface's lineage. When, later that night, the sea deposited the corpse on the beach, a corpse interrogation took place to which Marc Augé was witness. Many interpretations about his death were tried out through long discussions that reviewed Boniface's debts, illness, properties, clan and biography, until it was clear to all that one of Boniface's aunts had caused the death. She was the weak link in these long chains that tied Boniface to his fate, and my friend who had not obeyed his host's warning had, literally, nothing to do with his death. I had distributed causes and effects, attributed shame, guilt and responsibility, defined links between the people assembled on the beach, but the elders gathered around the wavering corpse had made an entirely different distribution, attribution and definition. As much anxiety, hate and anger, as much scepticism, scrutiny and belief circulated through the two networks, but the lines were not drawn in between the same points.

Free association 3: who kills the 40,000 people or so who die each year in car accidents in the United States? The cars? The road system? The Department of the Interior? No, drunk driving.[16] Who is responsible for this excess of alcohol? The wine merchants? The whisky manufacturers? The Department of Health? The Association of Bar Owners? No, the individual who drinks too much. Among all the possibilities only one is sociologically sustained: individuals who drink too much are the cause of most traffic accidents. This causal link is a premise, or a black box for all further reasonings in the matter. Once this is settled, there is controversy afterwards as to why the individual drivers drink too much. Are they *sick* people that should be cured and sent to a hospital, or *criminals* that should be punished and sent to prison? This depends on what definition of free will is given, on how the functioning of the brain is interpreted, on what force is granted to the law. Official spokespersons from sociology departments in universities, from voluntary associations, from the legal professions, from societies of physiologists, take positions and obtain figures proving the first or the second possibility. In defence of their positions they mobilise statistics, church teachings, common sense, repentant drinking drivers, principles of law, or brain neurology, anything that makes the claim such that if an opponent denies it, then they have to tackle its complex supports as well. As to the link between individual drinking and traffic violations, since no one disputes it, it is as straight and as necessary as the Alladians' attribution of Boniface's death to someone of his lineage.

The point I want to make with these 'free associations' is that they are in no way limited to certain kinds of people – that would limit anthropology to 'savage minds', to certain periods – that would limit anthropology to the study of our past – or to certain kinds of associations – that would limit anthropology to the study of world-views or of ideology. The same questions about causes, effects, links and spokespersons may be raised everywhere, thus opening an unlimited field of study for anthropology that can include Bulmer and his cassowary, the Karam and their kobtiy, Ostrom and his flying dinosaurs, Boniface's parents and their clan, the scallop larvae and their scientists, Gray and Bell and their networks, drinking drivers and their brains loaded with guilt and alcohol, Motabasi and his garden, Hutchins and his logical Trobrianders. We do not have to make any assumption about distorted world-views, nor do we have to assume that all these associations are equal, since they strive so much to tie heterogeneous elements together and to become unequal.

From the observer's point of view none of these people ever think either illogically or logically, but always sociologically; that is they go straight from elements to elements until a controversy starts. When this happens they look for stronger and more resistant allies, and in order to do so, they may end up mobilising the most heterogeneous and distant elements, thus mapping for themselves, for their opponents, and for the observers, what they value most, what they are most dearly attached to. 'Where thy treasure is, there will thy heart be also' (Luke, 12,34). The main difficulty in mapping the system of heterogeneous associations is in *not* making any additional assumption about how *real* they are. This does not mean that they are fictitious but simply that they resist certain trials – and that other trials could break them apart. A metaphor would help at this point to give the observer enough freedom to map the associations without distorting them into 'good' ones and 'bad' ones: sociologics are much like road maps; all paths go to some place, no matter if they are trails, tracks, highways or freeways, but they do not all go to the same place, do not carry the same traffic, do not cost the same price to open and to maintain. To call a claim 'absurd' or knowledge 'accurate' has no more meaning than to call a smuggler trail 'illogical' and a freeway 'logical'. The only things we want to know about these sociological pathways is where they lead to, how many people go along them with what sort of vehicles, and how easy they are to travel; not if they are wrong or right.

Part C
Who Needs Hard Facts?

In Part A we introduced symmetry between claims by distributing qualities equally among all the actors – openness, accuracy, logic, rationality – and defects – such as closure, fuzziness, absurdity, irrationality. Then, in Part B, we showed that this equal distribution did not stop any of the actors, when they dissented, from accusing some others of being 'grossly mistaken', 'inaccurate', 'absurd', and so on. To be sure, these accusations no longer told anything about the *form* of the claims which were attacked – since everyone is by now as logical as everyone else – but they nevertheless revealed by degrees the *content* of different associations clashing with one another.

In other words, all this business about rationality and irrationality is the result of an attack by someone on associations that stand in the way. They reveal the extent of a network and the conflict between what will stay inside and what will fall through its mesh. The important consequence is the same as that we drew at the end of Chapter 2 about the end of controversies: it is no use being relativist about claims which are not attacked; nature talks straight without any interference or bias, exactly like water flows regularly through a system of thousands of pipes, if there is no gap between them. This result may be extended to all claims: if they are not attacked, people know exactly what nature is; they are objective; they tell the truth; they do not live in a society or in a culture that could influence their grasp of things, they simply *grasp* things in themselves; their spokespersons are not 'interpreting' phenomena, nature talks through them directly. Insofar as they consider all the black boxes well sealed, people do not, any more than scientists, live in a world of fiction, representation, symbol, approximation, convention: they are simply *right*.

The question to raise then is when and why an attack that *crosses* someone else's path is possible, one that generates, at the intersection, the whole gamut of accusations (Part A), revealing step by step to what other unexpected elements a claim is tied (Part B). In other words we now have to get close to the clashes between the inside and the outside of the networks.

(1) Why not soft facts instead?

The first thing we have to understand is that the conditions for clashes between claims are not very often met. Let us take an example.

> 'An apple a day keeps the doctor away,' the mother said handing out a glowing red apple to her son, expecting a grin. 'Mother,' replied the child indignantly, 'three NIH studies have shown that on a sample of 458 Americans of all ages there was no statistically significant decrease in the number of house calls by family doctors; no, I will not eat this apple.'

What is out of step in this anecdote? The child's answer, for it mobilises too many elements in a situation that did not require it. What was expected? A smile, no reply, a quip, the repetition of the proverb, or, better, its completion ('An apple a day ....,' said the mother, 'keeps the doctor away,' answered the kid jokingly). Why does the intrusion of the National Institute of Health's statistics into the exchange seem so awkward? Because the son behaves as if he were in a controversy similar to the ones we studied in Chapter 1, fighting against his mother, expecting her to reply with more statistics, thus feeding back into the proof race! What did the mother expect instead? Not even a reply, nothing even vaguely related to a discussion, with proofs and counter-proofs. We will not understand anything about technoscience if we do not size up the vast distance between the son and the mother, between *harder facts* and *softer ones*.

At the beginning of Chapter 3, I presented the quandary of fact-builders. They have to enrol many others so that they participate in the continuing construction of the fact (by turning the claims into black boxes), but they also have to control each of these people so that they pass the claim along without transforming it either into some other claim or into someone else's claim. I said it was a difficult task, because each of the potential helping hands, instead of being a 'conductor,' may act in multifarious ways behaving as a 'multi-conductor': they may have no interest whatsoever in the claim, shunt it towards some unrelated topic, turn it into an artefact, transform it into something else, drop it altogether, attribute it to some other author, pass it along as it is, confirm it, and so on. As the reader may recall, the centrality of this process is the first principle of this book, on which everything else is built. The paradox of the fact-builders is that they have simultaneously to *increase* the number of people taking part in the action – so that the claim spreads – and to *decrease* the number of people taking part in the action – so that the claim spreads *as it is*. In Chapters 3 and 4 I followed in some detail cases where this paradox was solved by translating interests and tying them with non-human resources, thus producing machines and mechanisms. Having reached the last part of the present chapter we can now understand that these features of technoscience which are the rule inside the networks are the *exception* in between their meshes.

What then could the rule be? The claims will be at once transferred and transformed. Consider the proverb above; it has spread for many centuries from mouth to ear. Who is the author? This is unknown, it is common wisdom, no one cares, the question is meaningless. Is it objective, that is, does it refer to apples, health and doctors, or to the people who utter it? The question is meaningless, it never clashes with other claims – except in the anecdote above that for this reason appears queer. Is it wrong then? Not really, maybe, who cares? Then is it true? Probably, since it has been passed along for generations without a word of criticism. 'But, if it is true, why does it not stand the test of the son's counter-argument?', a rationalist could ask. Precisely, it has passed along so reliably for so long because everyone along the chain has *adapted* it to their own special context. At no point in the long history of this proverb has it been an argument fighting a counter-argument. It is not fit for use in a controversy between two strangers; it is only fit for reminding us, with a soft blow, which groups people who tell proverbs and their audience belong to – and in addition, it makes kids eat apples (and it is also possibly good for their health).

The son's practice of breaching modifies the *angle* at which claims encounter other claims and triggers irrationality as an effect of the clash. This breaching may be repeated with any of the innumerable instances offered by idle speech, twaddle, prattle, and chatter in bars, at parties, at home or at work. Every time a sentence like the proverb is answered by a counter-argument like that of the son, the same huge gap opens in the communication; friends, parents, lovers, buddies, party-goers become *estranged* at once, looking at each other with bewilderment. If in the bus your neighbour says, 'Nice weather today,eh?' and you answer 'That is a ridiculous statement, because the mean temperature today is four degrees below the normal average – computed on a hundred-year basis at Greenwich Observatory by Professor Collen and his colleagues using no less than fifty-five

weather stations. Check their methodology in *Acta Meteorologica*, you fool,'
your neighbour will think *you* are strange – and will probably move to another
seat. 'Nice weather today, eh?' is not a sentence fit for anything like what we have
seen so far in this book. Its *regime* of circulation, its way of passing from hand to
hand, the effects it generates seem vastly different from the statement we call
'scientific'. The breaching exercise repeats what has happened in the course of
recent history to many statements that were suddenly attacked by claims
circulating under a totally different regime. Most of what people say and used to
say is suddenly found wanting when considered from the inside of scientific
networks.

So maybe there is after all a radical difference between science and the rest, in
spite of what I have said in the two other parts?

(2) Hardening the facts

Yes, there is a difference, the breaching exercise indicates it clearly, but we have
to understand it without any additional divide. To grasp it we have to come back
to the first principle and to the quandary of the fact-builder. The simplest way to
spread a statement is to leave *a margin of negotiation* to each of the actors to
transform it as he or she sees fit and to *adapt it* to local circumstances. Then it will
be easier to interest more people in the claim since less control is exercised on
them. Thus, the statement will go from mouth to ear. However, there is a price to
pay for this solution. In such a venture the statement will be accommodated,
incorporated, negotiated, adopted and adapted by everyone and this will entail
several consequences:

first, the statement will be transformed by everyone but these transformations
will not be noticeable, because the success of the negotiation depends on the
absence of any comparison with the original statement;

second, it will have not one author but as many authors as there are members
along the chain;

third, it will not be a *new* statement, but will necessarily appear as an older one
since everyone will adapt it to their own past experience, taste and context;

fourth, even if the whole chain is changing opinion by adopting a new
statement – new, that is for the outside observer who behaves according to the
other regime below – this change will never be noticeable since there will be no
measurable baseline against which to notice the difference between older and
newer claims;

finally, since the negotiation is continuous along the chain and ignores clashes,
no matter how many resources are brought in to strengthen the claim, it will
always appear as a *softer* claim that does not break up the usual ways of
behaviour.[17]

Such is the regime under which the vast majority of claims circulate outside of
the new networks. It is one perfectly reasonable solution to the quandary of the

fact-builder, but it is one that produces only softer facts when compared with the second solution. This other solution to the quandary, as we saw in the previous chapters, is the one chosen by people who are called scientists and engineers. They prefer to increase control and to decrease the margin of negotiation. Instead of enrolling others by letting them transform the statement, they try to force them to take up the claim as it is. But as we have seen, there is a price to pay: few people may be interested, and many more resources have to be brought in to harden the facts. Consequently:

first, the statement may be transferred without being transformed – when everything works according to plan;

second, the owner of the original claim is designated – if he or she feels cheated, a bitter struggle ensues about who should be credited for the claim;

third, the claim is a *new* one that does not fit into the fabric of everyone's past experience – this is both a cause and a consequence of the diminishing margin of negotiation, and a cause and a consequence of the bitter fights for credit;

fourth, since each claim is measured by comparison with the previous ones, each new claim contrasts clearly with the background; thus it seems that a *historical* process is at work characterised by new beliefs that constantly shake the older ones;

and, finally, all the resources brought in to force people to assent are explicitly arrayed, making the claim a *harder* fact that appears to break through the usual softer ways of behaving and believing.

It is crucial to understand that these are two opposite solutions to the *same* paradox; 'harder' facts are not naturally better than 'softer' ones; they are the only solution if one wants to make others believe something uncommon. Nothing should be unduly added to these differences, even though some of the words used on the two lists seem to overlap with divides often used to oppose 'daily reasoning', 'savage mind', 'popular beliefs' and 'ancient and traditional sciences' to modern, civilised and scientific reasoning. In this argument, no assumption is made about minds or method. It is not assumed that the first solution provides closed, timeless, inaccurate, rigid and repetitive beliefs, whereas the second offers exact, hard and new knowledge. It is asserted simply that the same paradox may be solved in two different ways, one that extends long networks, the other that does not. If the first solution is chosen the fact-builder immediately *appears* as a stranger breaching what immediately *seem* to be old, timeless, stable and traditional ways. Irrationality is always an accusation made by someone building a network over someone else who stands in the way; thus, there is no divide between minds, but only shorter and longer networks. Harder facts are not the rule but the exception, since they are needed only in a very few cases to displace others on a large scale out of their usual ways. This will be our **fifth principle**.

It must be clear by now that it is impossible to say that everyone on earth should or could be a scientist at heart if only the forces of prejudice, superstition and passion could be overridden (see Part A). This proposition is as meaningless as saying that everyone of the 5 billion inhabitants of this planet ought to have a Rolls Royce. Hard facts are, by all means of assessing them, rare and costly occurrences that are only met in the few cases when someone tries to make others move out of their normal course and still wants them to participate faithfully in the enterprise. There is a direct relation between the *number* of people one wants to convince, the *angle* at which the claims clash with other claims and the *hardening* of the facts, that is the *number* of allies one has to fetch. Faced with harder facts we will no longer endow them with some innate and mysterious superiority, we will simply ask who is going to be attacked and displaced with them, relating the quality of the facts with the number of people moved out of their way, exactly as we could do when comparing a slingshot, a sword and an armoured tank or when comparing a small earth dam on a little brook with a huge concrete one on the Tennessee River.

(3) The sixth rule of method: just a question of scale...

At the end of this chapter we are now in a position to understand the many differences triggered by the accusation process between so-called 'traditional' cultures – that is the ones that are accused of *believing* in things – and the narrow scientific networks that, in order to grow everywhere, have to discover that all the statements used so far by people are weak, inaccurate, soft or wrong. To do so we have simply to follow the scientists in their work.

In order to strengthen their claims some of them have to go out of *their* way and *come back* with new *unexpected* resources so as to win the encounters they have at home with the people they wish to convince. What will happen during such a move? The traveller will *cross* the paths of many other people. We know from Parts A and B that it is this crossing that is going to trigger the accusations of irrationality. At every intersection new and unexpected associations between things, words, mores and people are revealed. However, this is not enough yet to generate huge differences between cultures. Pirates, merchants, soldiers, diplomats, missionaries, adventurers of all sorts have for centuries travelled through the world and got used to the diversity of cultures, religion and belief systems.

But consider the peculiar nature of crossing someone else's path when harder facts are at stake. Consider Bulmer sent to New Guinea, or Evans-Pritchard to Africa, or Hutchins to the Trobriand Islands. Consider the paleontologists trekking through the Nevada desert looking for fossile bones. Consider the geographers sent away to map out the Pacific Coast. Consider the botanists mandated to bring back all sorts of plants, fruits and herbs. Are these travellers interested in the people, the landscape, the customs, the forests, the oceans they go through? In a sense yes, because they want to use them in order to come back with more resources. In another sense, no, because they do not wish to settle in all these foreign places. If Bulmer goes away and stays for ever in New Guinea, becoming one of the Karam, his trip is wasted as far as hard facts are concerned. But if he comes back empty-handed without any information that can be used in theses or in articles to make his points, his whole trip is wasted as well, no matter how much he learned, understood and suffered. Since all these travellers are 'interested', they are going to learn everything they can along the road; but since they are not interested in remaining at any place in particular, only in coming back home, they are going to be sceptical about all the stories they are told. Because of this paradox the drama of the *Great Divide* unfolds. By the Great Divide is meant the summary of all the accusation processes that are made from within scientific networks against their outside. The sociologics of all the people crossed by these peculiar travellers sent away in order to come back is going to appear *by comparison* 'local', 'closed', 'stable', 'culturally determined'. Once the movement of the observer is deleted from the picture, it seems that there is an absolute divide between, on the one hand, all the cultures that 'believe' in things, and on the other hand, the one culture, ours, that 'knows' things (or will soon know them), between 'Them' and 'Us'.

Belief by rationalists in the existence of the Great Divide, as well as the denial of its existence by relativists, both depend on forgetting the *movement* of the observer moving away from home to come back heavily armed in order to strengthen the facts. The complete misunderstanding of the qualities and defects of Them and Us is sketched in Figure 5.4. As soon as the accuser's movement is put back into the picture, a difference appears, but it has nothing to do with a divide between belief and knowledge. It has simply to do with the *scale* at which the enrolling and controlling of people occurs.

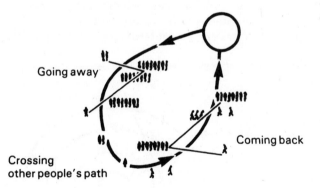

Figure 5.4

Can we say, for instance, that scientists moving through the world are more 'disinterested', more 'rational', more concerned by the things 'themselves', less 'culturally determined', more 'conscious' than the people they meet along the way? In a sense, yes, certainly, they are less interested in maintaining the societies they cross than members of these societies themselves! Thus, they are going to keep their distance, to be cooler, to be disbelievers. But in another sense they are as interested as everyone else in maintaining *their* own society back home – and this is why they so much wish to enrich science with one more piece of accurate information. If they were totally disinterested they would not take any notes, they would just loiter around, stay a few years, move away, and never come back.

All the conditions of a major misunderstanding are now filled. Because he is so *interested*, Bulmer, for example, is going to be maddeningly obsessed with his notebooks, double-checking all information, filling crates with materials, gathering all he can before running. As far as the Karam's belief in classification is concerned, Bulmer is cool as a cucumber, 'seeing through' their foreign solutions the influence of their local culture; but as far as Bulmer's belief in anthropology is concerned, the Karam are very cool indeed, seeing through his obsession for notes and accurate information the influence of the foreign culture he so dearly wishes to maintain and expand. 'Disinterested fanatics', such as Bulmer, are going to transform all the claims of all the people they meet into 'beliefs about' the world that require a special explanation. Bulmer cannot believe the Karam are right since he is not going to stay with them for ever; but he cannot be tolerant either and choose a sort of soft relativism that would not care a bit about what other people think, since he has to come back with a report on the Karam's belief system. So he is going to come back to his department *with their written beliefs* in taxonomy.[18] Once in New Zealand, the Karam's taxonomy will be compared with *all* the taxonomies brought back by other anthropologists. At this point the misunderstanding is completed: the Karam will be said to have only one way of looking at the world, the anthropologists to have many. The Karam's peculiar way of choosing among classificatory patterns begs an explanation that will be found in their society; the anthropologists' views that cover all the

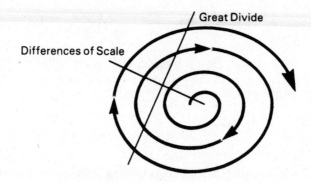

Figure 5.5

patterns are not to be explained by their society: they are the right way. They will call *ethno*zoology the belief systems of the local Karam and zoology the knowledge of the universal scientific network. Although each sociologic is building its world by incorporating birds, plants, rocks, together with people, it will appear, at the end of many trips abroad, that only 'They' have an *anthropomorphic* belief system, whereas 'We' have a disinterested outlook on the world only slightly biased by our 'culture'. In Figure 5.5 I have sketched two possible renderings of the differences: the first one is obtained by tracing *a* Divide between Them and Us; the second by measuring *many* variations in the size of the networks. The Great Divide makes the supposition that there is, on the right hand, knowledge embedded in society, and, on the left hand, knowledge independent of society. We make no such supposition. The general fusion of knowledge and society is the same in all cases – a spiral in the diagram – but the length of the curve varies from one to the other.

'Interest' and 'disinterestedness' are words like 'rational' or 'irrational'; they are meaningless as long as we do not consider the movement of the scientist through the world. This will constitute our **sixth rule of method:** when faced with an accusation of irrationality, or simply with beliefs in something, we will never believe that people believe in things or are irrational, we will never look for which rule of logic has been broken, we will simply consider the angle, direction, movement and *scale* of the observer's displacement.

Of course, now that we are freed from all these debates about 'rationality', 'relativism', 'culture', and the extent of the Great Divide, we have one more question to tackle, the most difficult of all: where does the difference of scale come from?

CHAPTER 6

Centres of Calculation

Prologue
The domestication of the savage mind

At dawn, 17 July 1787, Lapérouse, captain of *L'Astrolabe*, landed at an unknown part of the East Pacific, on an area of land that was called 'Segalien' or 'Sakhalin' in the older travel books he had brought with him. Was this land a peninsula or an island? He did not know, that is no one in Versailles at the court of Louis XVI, no one in London, no one in Amsterdam in the headquarters of the West Indies Company, could look at a map of the Pacific Ocean and decide whether the engraved shape of what was called 'Sakhalin' was tied to Asia or was separated by a strait. Some maps showed a peninsula, others showed an island; and a fierce dispute had ensued among European geographers as to how accurate and credible the travels books were and how precise the reconnaissances had been. It is in part because there were so many of these disputes – similar to the profusion we studied in Part I – on so many aspects of the Pacific Ocean, that the king had commissioned Lapérouse, equipped two ships, and ordered him to draw a complete map of the Pacific.[1]

The two ships had been provided, as scientific satellites are today, with all the available scientific instruments and skill; they were given better clocks to keep the time, and thus measure the longitude more accurately; they were given compasses to measure the latitude; astronomers had been enlisted to mend and tend the clocks and to man the instruments; botanists, mineralogists and naturalists were on board to gather specimens; artists had been recruited to sketch and paint pictures of those of the specimens that were too heavy or too fragile to survive the return trip; all the books and travel accounts that had been written on the Pacific had been stocked in the ship's library to see how they compared with what the travellers would see; the two ships had been loaded with goods and bargaining chips in order to evaluate all over the world the relative prices of gold, silver, pelts, fish, stones, swords, anything that could be bought and sold at a profit, thus trying out possible commercial routes for French shipping.

This morning in July, Lapérouse was very surprised and pleased. The few savages – all males – that had stayed on the beach and exchanged salmon for pieces of iron were much less 'savage' than many he had seen in his two years of travel. Not only did they seem to be sure that Sakhalin was an island, but they also appeared to understand the navigators' interest in this question and what it was to draw a map of the land viewed from above. An older Chinese sketched on the sand the country of the 'Mantchéoux', that is, China, and his island; then he indicated with gestures the size of the strait separating the two. The scale of the map was uncertain, though, and the rising tide soon threatened to erase the precious drawing. So, a younger Chinese took up Lapérouse's notebook and pencil and drew another map noting the scale by little marks, each signifying a day of travel by canoe. They were less successful in indicating the scale for the depth of the strait; since the Chinese had little notion of the ship's draught, the navigators could not decide if the islanders were talking of relative or of absolute size. Because of this uncertainty, Lapérouse, after having thanked and rewarded these most helpful informants, decided to leave the next morning and to sight the strait for himself, and, hopefully, to cross it and reach Kamchatka. The fog, adverse winds and bad weather made this sighting impossible. Many months later, when they finally reached Kamchatka, they had not seen the strait, but relied on the Chinese to decide that Sakhalin was indeed an island. De Lesseps, a young officer, was asked by Lapérouse to carry the maps, the notebooks and the astronomical bearings they had gathered for two years back to Versailles. De Lesseps made the trip on foot and on horseback under the protection of the Russians, carrying with him these precious little notebooks; one entry among thousands in the notebooks indicated that the question of the Sakhalin island was settled and what the probable bearing of the strait was.

This is the kind of episode that could have been put to use, at the beginning of Chapter 5, in order to make the Great Divide manifest. At first sight, it seems that the differences between Lapérouse's enterprise and those of the natives is so colossal as to justify a deep distinction in cognitive abilities. In less than three centuries of travels such as this one, the nascent science of geography has gathered more knowledge about the shape of the world than had come in millennia. The *implicit* geography of the natives is made *explicit* by geographers; the *local* knowledge of the savages becomes the *universal* knowledge of the cartographers; the fuzzy, approximate and ungrounded *beliefs* of the locals are turned into a precise, certain and justified *knowledge*. To the partisans of the Great Divide, it seems that going from ethnogeography to geography is like going from childhood to adulthood, from passion to reason, from savagery to civilisation, or from first degree intuitions to second degree reflexion.

However, as soon as we apply the sixth rule of method, the Great Divide disappears and other little differences become visible. As I showed in the last chapter, this rule asks us not to take a position on rationality, but simply to consider the movement of the observer, its angle, direction and scale.

Lapérouse crosses the path of the Chinese fishermen *at right angles*; they have never seen each other before and the huge ships are not here to settle. The Chinese have lived here for as long as one can remember whereas the French fleet remains with them for a day. These families of Chinese, as far as one can tell, will remain around for years, maybe centuries; *L'Astrolabe* and *La Boussole* have to reach Russia before the end of the summer. In spite of this short delay, Lapérouse does not simply cross the path of the Chinese ignoring the people on shore. On the contrary, he learns from them as much as he can, describing their culture, politics and economics – after one day of observation! – sending his naturalists all over the forest to gather specimens, scribble notes, take the bearings of stars and planets. Why are they all in a hurry? If they were interested in the island could they not stay longer? No, because they are not so much interested in this place as they are in bringing this place *back* first to their ship, and second to Versailles.

But they are not only in a hurry, they are also under enormous pressure to gather traces that have to be of a certain quality. Why is it not enough to bring back to France personal diaries, souvenirs and trophies? Why are they all so hard-pressed to take precise notes, to obtain and double-check vocabularies from their informants, to stay awake late at night writing down everything they have heard and seen, labelling their specimens, checking for the thousandth time the running of their astronomical clocks? Why don't they relax, enjoy the sun and the tender flesh of the salmon they catch so easily and cook on the beach? Because the people who sent them away are not so much interested in their coming back as they are in the possibility of sending *other* fleets *later*. If Lapérouse succeeds in his mission, the next ship will know if Sakhalin is a peninsula or an island, how deep the strait is, what the dominant winds are, what the mores, resources and culture of the natives are *before* sighting the land. On 17 July 1787, Lapérouse is *weaker* than his informants; he does not know the shape of the land, does not know where to go; he is at the mercy of his guides. Ten years later, on 5 November 1797 the English ship *Neptune* on landing again at the same bay will be much *stronger* than the natives since they will have on board maps, descriptions, log books, nautical instructions – which to begin with will allow them to know that this is the 'same' bay. For the new navigator entering the bay, the most important features of the land will all be seen for the *second* time – the first time was when reading in London Lapérouse's notebooks and considering the maps engraved from the bearings De Lesseps brought back to Versailles.

What will happen if Lapérouse's mission does not succeed? If De Lesseps is killed and his precious treasure scattered somewhere on the Siberian tundra? Or if some spring in the nautical clocks went wrong, making most of the longitudes unreliable? The expedition is wasted. For many more years a point on the map at the Admiralty will remain controversial. The next ship sent away will be *as weak* as *L'Astrolabe*, sighting the Segalien (or is it Sakhalin?) island (or is it a peninsula?) for the *first* time, looking again for native informants and guides; the divide will remain as it is, quite small since the frail and uncertain crew of the

Neptuna will have to rely on natives as poor and frail as them. On the other hand, if the mission succeeds, what was at first a small divide between the European navigator and the Chinese fishermen will have become larger and deeper since the *Neptuna* crew will have less to learn from the natives. Although there is at the beginning not much difference between the abilities of the French and the Chinese navigators, the difference will grow if Lapérouse is part of a network through which the ethnogeography of the Pacific is accumulated in Europe. An asymmetry will slowly begin to take shape between the 'local' Chinese and the 'moving' geographer. The Chinese will remain savage (to the European) and as strong as the *Neptuna* crew, if Lapérouse's notebooks do not reach Versailles. If they do, the *Neptuna* will be better able to *domesticate* the Chinese since everything of their land, culture, language and resources will be known on board the English ship before anyone says a word. Relative degrees of savagery and domestication are obtained by many little tools that make the wilderness known in advance, predictable.

Nothing reveals more clearly the ways in which the two groups of navigators talk at cross purposes, so to speak, than their interest in the inscription. The accumulation that will generate an asymmetry hinges upon the possibility for some traces of the travel to go back to the place that sent the expedition away. This is why the officers are all so much obsessed by bearings, clocks, diaries, labels, dictionaries, specimens, herbaries. Everything depends on them: *L'Astrolabe* can sink provided the inscriptions survive and reach Versailles. This ship travelling through the Pacific is an instrument according to the definition given in Chapter 2. The Chinese, on the other hand, are not all that interested in maps and inscriptions – not because they are unable to draw them (on the contrary their abilities surprise Lapérouse very much) but simply because the inscriptions are not the *final goal* of their travel. The drawings are no more than *intermediaries* for their exchanges between themselves, intermediaries which are used up in the exchange and are not considered important in themselves. The fishermen are able to generate these inscriptions at will on any surface like sand or even on paper when they meet someone stupid enough to spend only a day in Sakhalin who nevertheless wishes to know everything fast for some other unknown foreigner to come back later and safer. There is no point in adding any cognitive difference between the Chinese navigators and the French ones; the misunderstanding between them is as complete as between the mother and the child in Chapter 5 and for the same reason: what is an intermediary of no relevance has become the beginning and the end of a cycle of capitalisation. The difference in their movement is enough and the different emphasis they put on inscriptions ensues. The map drawn on sand is worthless for the Chinese who do not care that the tide will erase it; it is a treasure for Lapérouse, his main treasure. Twice, in his long travels, the captain was fortunate enough to find a faithful messenger who brought his notes back home. De Lesseps was the first; Captain Phillip, met at Botany Bay in Australia in January 1788, was the second. There was no third time. The two ships disappeared and the only traces that were found, well into the nineteenth century, were not maps and herbariums, but the hilt of a sword and a piece of the stern with a fleur-de-lis on it, that had become the door of a savage's hut. On the third leg of their journey the French navigators had not been able to domesticate the savage lands and peoples; consequently, nothing is known with certainty about this part of their voyage.

Part A
Action at a distance

(1) Cycles of accumulation

Can we say that the Chinese sailors Lapérouse met did not know the shape of their coasts? No, they knew it very well; they had to since they were born there. Can we say that these Chinese did not know the shape of the Atlantic, of the Channel, of the river Seine, of the park of Versailles? Yes, we are allowed to say this, they had no idea of them and probably they could not care less. Can we say that Lapérouse knew this part of Sakhalin before landing there? No, it was his first encounter with it, he had to fumble in darkness, taking soundings along the coast. Are we allowed to say that the crew of the *Neptuna* knew this coast? Yes, we may say this, they could look at Lapérouse's notes, and compare his drawings of the landings with what they saw themselves; less sounding, less fumbling in the dark. Thus, the knowledge that the Chinese fishermen had and that Lapérouse did *not* possess had, in some still mysterious way, been provided to the crew of the English ship. So, thanks to this little vignette, we might be able to define the word knowledge.

The first time we encounter some event, we do not know it; we start knowing something when it is at least the *second* time we encounter it, that is, when it is familiar to us. Someone is said to be knowledgeable when whatever happens is only one instance of other events already mastered, one member of the same family. However, this definition is too general and gives too much of an advantage to the Chinese fishermen. Not only have they seen Sakhalin twice, but hundreds and even thousands of times for the more elderly. So they will always be more knowledgeable than these white, ill-shaven, capricious foreigners who arrive at dawn and leave at dusk. The foreigners will die en route, wrecked by typhoons, betrayed by guides, destroyed by some Spanish or Portuguese ship, killed by yellow fever, or simply eaten up by some greedy cannibals . . . as probably happened to Lapérouse. In other words, the foreigner will always be weaker than any one of the peoples, of the lands, of the climates, of the reefs, he meets around the world, always at their mercy. Those who go away from the lands in which they are born and who cross the paths of other people disappear without trace. In this case, there is not even time for a Great Divide to be drawn; no accusation process takes place, no trial of strength between different sociologics occurs, since the moving element in this game, that is the foreigner, vanishes at the first encounter.

If we define knowledge as familiarity with events, places and people seen many times over, then the foreigner will always be the weakest of all except if, by some extraordinary means, whatever happens to him happens at least twice; if the islands he has never landed at before have already been seen and carefully studied, as was the case with the navigator of the *Neptuna*, then, and only then, the moving foreigner might become stronger than the local people. What could these 'extraordinary means' be? We know from the Prologue that it is not enough for a foreigner to have been preceded by one, or two, or hundreds of others, as long as these predecessors either have vanished without trace, or have come back with obscure tales, or keep for themselves rutters only *they* can read, because, in these three cases, the new sailor has gained nothing from his predecessors' travels; for him, everything will happen the first time. No, he will gain an edge only if the other navigators have found a way to *bring* the lands *back with them* in such a manner that he will *see* Sakhalin island, for the first time, at leisure, in his own home, or in the Admiralty office, while smoking his pipe . . ..

As we see, what is called 'knowledge' cannot be defined without understanding what *gaining* knowledge means. In other words, 'knowledge' is not something that could be described by itself or by opposition to 'ignorance' or to 'belief', but only by considering a whole cycle of accumulation: how to bring things back to a place for someone to see it for the first time so that others might be sent again to bring other things back. How to be familiar with things, people and events, which are *distant*. In Figure 6.1 I have sketched the same movement as in Figure 5.4 but instead of focusing on the accusation that takes place at the intersection, I have focused on the accumulation process.

Expedition number one disappears without trace, so there is no difference in 'knowledge' between the first and the second that fumbles its way in darkness always at the mercy of each of the people whose path is crossed. More fortunate

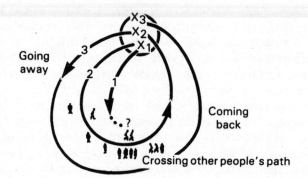

Figure 6.1

than the first, this second expedition not only comes back but brings something (noted X2 in the drawing) that allows the third to be so familiar with the coastline that they can quickly move to other lands bringing home parts of a map of a *new* territory (X3). At every run of this accumulation cycle, more elements are gathered in the centre (represented by a circle at the top); at every run the asymmetry (at the bottom) between the foreigners and the natives grows, ending today in something that indeed looks like a Great Divide, or at least like a disproportionate relation between those equipped with satellites who localise the 'locals' on their computer maps without even leaving their air-conditioned room in Houston, and the helpless natives who do not even see the satellites passing over their heads.

We should not be in a rush to decide what are these 'extraordinary means', what these things noted 'X' in the drawing are, which are brought back by the navigators. We first have to understand under what conditions a navigator can sail overseas and come *back*, that is how a cycle may be drawn at all. To do this, we have to take a much earlier example when these travels abroad were yet more perilous. Three centuries before Lapérouse, in 1484, King John II of Portugal convened a small scientific commission to help navigators finding their way to the Indies.[2]

At this time a first condition has been fulfilled: the heavy and sturdy carracks designed by the Portuguese did not disintegrate any more in storms or long sojourns at sea; the wood of which they were built and the way they were careened made them stronger than waves and tides. In the definition of the term I gave in Chapter 3, they acted as *one element*; they had become a clever machination to control the many forces that tried out their resistance. For instance, all sorts of wind directions, instead of slowing the ships down, were turned into allies by a unique combination of lateen and square rigs. This combination allowed a smaller crew to man a bigger ship, which made crew members less vulnerable to malnutrition and plagues, and captains less vulnerable to mutinies. The bigger size of the carracks made it possible to embark bigger guns which, in turn, rendered more predictable the outcome of all military encounters with the many tiny pirogues of the natives. This size also rendered it practical to bring back a bigger cargo (if there were a return trip).

When the scientific commission convened, the carracks were already very mobile and versatile tools, able to extract compliance from the waves, the winds, the crew, the guns and the natives, but not yet from the reefs and the coastline. These were always more powerful than the carracks since they appeared unexpectedly, wrecking the ships one after the other. How to localise in advance all the rocks instead of being, so to speak, *localised* by them without warning? The solution of the commission was to use the furthest-fetched of all possible helping hands, the sun and the stars, whose slow declination could be turned, with the help of instruments to determine angles, of tables to make the calculation, of training to prepare the pilots, into a not-too-inaccurate approximation of the latitude. After years of compilation, the commission wrote the *Regimento do Astrolabio and do Qadrante.* This book on board every ship gave very practical directions on how to use the quadrant and how to measure the latitude by entering the date, the time, the angle of the sun with the horizon; in addition, the commission compiled all the bearings of good quality that had been made at various latitudes, systematically adding each reliable one. Before this commission, capes, reefs and shoals were stronger than all the ships, but after this, the carracks plus the commission, plus the quadrants, plus the sun, had tipped the balance of forces in favour of the Portuguese carracks: the dangerous coastline could not rear up treacherously and interrupt the movement of the ship.

Still, even with the winds, the wood, the coastline, the crews, the sun, disciplined, aligned, well-drilled and clearly on King John's side, there is no guarantee that a cycle of accumulation will be drawn that will *start* from him and *end* with him, in Lisbon. For instance, Spanish ships may divert the carracks out of their way; or the captains with their ships loaded with precious spices may betray the king and sell them elsewhere to their profit; or Lisbon's investors might keep for themselves most of the profit and baulk at equipping a new fleet to continue the cycle. Thus, in addition to all his efforts in ship designing, cartography and nautical instructions, the king must invent many new ways to extract compliance from investors, captains, custom officers; he must insist on legal contracts to bind, as much as he can, with signatures, witnesses and solemn oaths, his pilots and admirals; he must be adamant on well-kept accounting books, on new schemes to raise money and to share benefits; he must insist on each log book being carefully written, kept out of the enemy's sight, and brought back to his offices in order for its information to be compiled.

Together with the Prologue, this example introduces us to the most difficult stage of this long travel that leads us not through the oceans, but through technoscience. This cumulative character of science is what has always struck scientists and epistemologists most. But in order to grasp this feature, we have to keep in view all the conditions that allow a cycle of accumulation to take place. At this point the difficulties seem enormous because these conditions *cut across* divisions usually made between economic history, history of science, history of technology, politics, administration or law, since the cycle drawn by King John may leak at any seam: it may be that a legal contract is voided by a court, or a shifting political alliance gives Spain the upper hand, or the timber of a ship does not resist a typhoon, or a miscalculation in the *Regiment* sends a fleet ashore, or a mistake in the appraisal of a price renders a purchase worthless, or a microbe brings the plague back with the spices . . .. There is no way to neatly order these links into categories, since they have all been woven together, like the many threads of a macramé, to make up for one another's weaknesses. All the distinctions one could wish to make between domains (economics, politics, science, technology, law) are less important than the unique movement that makes all of these domains conspire towards the same goal: a cycle of accumulation that allows a point to become a *centre* by acting at a distance on many other points.

If we wish to complete our journey we have to define words that help us to follow this heterogeneous mixture and not to be interrupted and baffled every time the cycle-builders change gears going from one domain into another. Will we call 'knowledge' what is accumulated at the centre? Obviously, it would be a bad choice of words because becoming familiar with distant events requires, in the above examples, kings, offices, sailors, timber, lateen rigs, spice trades, a whole bunch of things not usually included in 'knowledge'. Will we call it 'power' then? That would also be a mistake because the reckoning of lands, the filling-in of log books, the tarring of the careen, the rigging of a mast, cannot without absurdity be put under the heading of this word. Maybe we should speak of 'money' or more abstractly of 'profit' since this is what the cycle adds up to. Again, it would be a bad choice because there is no way to call profit the small bundle of figures De Lesseps brings back to Versailles or the rutters put in the hands of King John; nor is the profit the main inducement for Lapérouse, his naturalists, his geographers and his linguists. So how are we to call what is brought back? We could of course talk of 'capital' that is something (money, knowledge, credit, power) that has no other function but to be instantly reinvested into another cycle of accumulation. This would not be a bad word, especially since it comes from *caput*, the head, the master, the centre, the capital of a country, and this is indeed a characterisation of Lisbon, Versailles, of all the places able to join the beginning and the end of such a cycle. However, using this expression would be begging the question: what is capitalised is necessarily turned into capital, it does not tell us what it is – besides, the word 'capitalism' has had too confusing a career . . ..

No, we need to get rid of all categories like those of power, knowledge, profit or capital, because they divide up a cloth that we want seamless in order to study it as we choose. Fortunately, once we are freed from the confusion introduced by all these traditional terms the question is rather simple: how to act at a distance on unfamiliar events, places and people? Answer: by *somehow* bringing home these events, places and people. How can this be achieved, since they are distant? By inventing means that (a) render them *mobile* so that they can be brought back; (b) keep them *stable* so that they can be moved back and forth without additional distortion, corruption or decay, and (c) are *combinable* so that whatever stuff they are made of, they can be cumulated, aggregated, or shuffled like a pack of cards. If those conditions are met, then a small provincial town, or an obscure laboratory, or a puny little company in a garage, that were at first as weak as any other place will become centres dominating at a distance many other places.

(2) The mobilisation of the worlds

Let us now consider some of the means that allow mobility, stability or combinability to improve, making domination at a distance feasible. Cartography is such a dramatic example that I chose it to introduce the argument. There is no way to bring the lands themselves to Europe, nor is it possible to gather in Lisbon or at Versailles thousands of native pilots telling navigators where to go and what to do in their many languages. On the other hand, all the voyages are wasted if nothing except tales and trophies comes back. One of the 'extraordinary means' that have to be devised is to use travelling ships as so many instruments, that is as *tracers* that draw on a piece of paper the shape of the encountered land. To obtain this result, one should discipline the captains so that, whatever happens to them, they take their bearings, describe the shoals, and send them back. Even this is not enough, though, because the centre that gathers all these notebooks, written differently according to different times and places of entry, will produce on the drafted maps a chaos of conflicting shapes that even experienced captains and pilots will hardly be able to interpret. In consequence, many more elements have to be put on board the ships so that they can calibrate and discipline the extraction of latitudes and longitudes (marine clocks, quadrants, sextants, experts, preprinted log books, earlier maps). The travelling ships become costly instruments but what they bring or send back can be transcribed on the chart almost immediately. By coding every sighting of any land in longitude and latitude (two figures) and by sending this code back, the shape of the sighted lands may be redrawn by those who have not sighted them. We understand now the crucial importance of these bundles of figures carried around the world by De Lesseps and the skipper of the *Neptuna*, Captain Martin: they were some of these stable, mobile and combinable elements that allow a centre to dominate faraway lands.

At this point those who were the weakest because they remained at the centre and saw nothing start becoming the strongest, familiar with *more* places not only than any native but than any travelling captain as well; a 'Copernican revolution' has taken place. This expression was coined by the philosopher Kant to describe what happens when an ancient discipline, uncertain and shaky until then, becomes cumulative and 'enters the sure path of a science'. Instead of the mind of the scientists revolving around the things, Kant explains, the things are made to revolve around the mind, hence a revolution as radical as the one Copernicus is said to have triggered. Instead of being dominated by the natives and by nature, like the unfortunate Lapérouse staking his life every day, the cartographers in Europe start gathering in their chart rooms – the most important and costliest of all laboratories until the end of the eighteenth century – the bearings of all lands. How large has the earth become in their chart rooms? No bigger than an *atlas* the plates of which may be flattened, combined, reshuffled, superimposed, redrawn at will. What is the consequence of this change of scale? The cartographer *dominates* the world that dominated Lapérouse. The balance of forces between the scientists and the earth has been reversed; cartography has entered the sure path of a science; a centre (Europe) has been constituted that begins to make the rest of the world turn around itself.

One other way of bringing about the same Copernican revolution is to gather *collections*. The shapes of the lands have to be coded and drawn in order to become mobile, but this is not the case for rocks, birds, plants, artefacts, works of art. Those can be extracted from their context and taken away during *expeditions*. Thus the history of science is in large part the history of the mobilisation of anything that can be made to move and shipped back home for this universal census. The outcome, however, is that in many instances stability becomes a problem because many of these elements die – like the 'happy savages' anthropologists never tired of sending to Europe: or become full of maggots – like grizzly bears zoologists have stuffed too quickly; or dry up – like precious grains naturalists have potted in too poor a soil. Even those elements which can withstand the trip, like fossils, rocks or skeletons, may become meaningless once in the basement of the few museums that are being built in the centres, because not enough context is attached to them. Thus, many inventions have to be made to enhance the mobility, stability and combinability of collected items. Many instructions are to be given to those sent around the world on how to stuff animals, how to dry up plants, how to label all specimens, how to name them, how to pin down butterflies, how to paint drawings of the animals and trees no one can yet bring back or domesticate. When this is done, when large collections are initiated and maintained, then again the same revolution occurs. The zoologists in their Natural History Museums, without travelling more than a few hundred metres and opening more than a few dozen drawers, travel through all the continents, climates and periods. They do not have to risk their life in these new Noah's Arks, they only suffer from the dust and stains made by plaster of Paris. How could one be surprised if they start to *dominate* the ethnozoology of all the other peoples? It is the contrary that would indeed be surprising. Many common features that could not be visible between dangerous animals far away in space and time can easily appear between one case and the next! The zoologists *see new* things, since this is the first time that so many creatures are drawn together in front of someone's eyes; that's all there is in this mysterious beginning of a science. As I said in Chapter 5, it is simply a question of scale. It is not at the cognitive differences that we should marvel, but at this general mobilisation of the world that endows a few scientists in frock coats, somewhere in Kew Gardens, with the ability to visually dominate all the plants of the earth.[3]

There is no reason, however, to limit the mobilisation of stable and combinable traces to those places where a human being can go in the flesh during an expedition. *Probes* may be sent instead. For instance, the people who dig an oil rig would very much like to know how many barrels of oil they have under their feet. But there is no way to go inside the ground and to see it. This is why, in the early 1920s, Conrad Schlumberger, a French engineer, had the idea of sending an electric current through the soil to measure the electrical resistance of the layers of rocks at various places.[4] At first, the signals carried confusing shapes back to their sender, as confusing as the first rutters brought back to the early cartographers. The signals were stable enough, however, to later allow the geologists to go *back and forth* from the new electric maps to the charts of the sediments they had drawn earlier. Instead of simply digging oil out, it became possible to accumulate traces on maps that, in turn, allowed engineers to direct the exploration less blindly. An accumulation cycle was started where oil, money, physics and geology helped accumulate one another. In a few decades, dozens of different instruments were devised and stacked together, slowly transforming the invisible and inaccessible reserves into loggings a few men could dominate by sight. Today, every derrick is used not only to pump oil but to carry sensors of all sorts deep inside the ground. At the surface, the *Schlumberger* engineers, in a movable lorry full of computers, are reading the results of all these measurements inscribed on millimetred paper hundreds of feet long.

The main advantage of this logging is not only in the mobility it provides to the deep structure of the ground, not only in the stable relations it establishes between a map and this structure, but in the *combinations* it allows. There is at first no simple connection between money, barrels, oil, resistance, heat; no simple way of tying together a banker in Wall Street, an exploration manager at Exxon headquarters, an electronician specialised in weak signals at Clamart near Paris, a geophysicist in Ridgefield. All these elements seem to pertain to different realms of reality: economics, physics, technology, computer science. If instead we consider the cycle of accumulation of stable and combinable mobiles, we literally *see* how they can go together. Consider, for instance, the 'quick look logging' on an oil platform in the North Sea: all the readings are first coded in binary signals and stocked for future, more elaborate calculations, then they are reinterpreted and redrawn on computers which spew out of the printers logs which are not scaled in ohms, microseconds or microelectrovolts, but directly in number of barrels of oil. At this point, it is not difficult to understand how platform managers can plan their production curve, how economists can add to these maps a few calculations of their own, how the bankers may later use these charts to evaluate the worth of the company, how they can all be archived to help the government calculate the proven reserves, a very controversial issue. Many things can be done with this paper world that cannot be done with the world.

For a Copernican revolution to take place it does not matter what means are used provided this goal is achieved: a shift in what counts as centre and what counts as periphery. For instance, nothing dominates us more than the stars. It seems that there is no way to reverse the scale and to make us, the astronomers, able to master the sky above our heads. The situation is quickly reversed, however, when Tycho Brahe, inside a well equipped *observatory* built for him at Oranenbourg, starts not only to write down on the same homogeneous charts the positions of the planets, but also to gather the sightings made by other astronomers all over Europe which he had asked them to write down on the same preprinted forms he has sent them.[5] Here again a virtuous cumulative circle starts to unfold if all sightings at different places and times are gathered together and synoptically displayed. The positive loop runs all the more rapidly, if the same Brahe is able to gather in the same place not only fresh observations made by him and his colleagues, but all the older books of astronomy that the printing press has made available at a low cost. His mind has not undergone a mutation; his eyes are not suddenly freed from old prejudices; he is not watching the summer sky more carefully than anyone before. But he is the first indeed to consider at a glance the summer sky, plus his observations, plus those of his collaborators, plus Copernicus' books, plus many versions of Ptolemy's *Almagest*; the first to sit at the beginning and at the end of a long network that generates what I will call **immutable and combinable mobiles**. All these charts, tables and trajectories are conveniently at hand and combinable at will, no matter whether they are twenty centuries old or a day old; each of them brings celestial bodies billions of tons heavy and hundreds of thousands of miles away to the size of a point on a piece of paper. Should we be surprised then if Tycho Brahe pushes astronomy further on 'the sure path of a science'? No, but we should marvel at those many humble means that turn stars and planets into pieces of paper inside the observatories that soon will be built everywhere in Europe.

The task of dominating the earth or the sky is almost equalled in difficulty by that of dominating a country's economy. There is no telescope to see it, no collection to gather it, no expedition to map it out. Here again in the case of economics, the history of a science is that of the many clever means to transform whatever people do, sell and buy into something that can be mobilised, gathered, archived, coded, recalculated and displayed. One such means is to launch *enquiries* by sending throughout the country pollsters, each with the same predetermined questionnaire that is to be filled in, asking managers the same questions about their firms, their losses and profits, their predictions on the future health of the economy. Then, once all the answers are gathered, other tables may be filled in that summarise, reassemble, simplify and rank the firms of a nation. Someone looking at the final charts is, in some way, considering the economy. Of course, as we know from earlier chapters, controversies will start about the accuracy of these charts and about who may be said to speak in the name of the economy. But as we also know, other graphic elements will be fed back in the controversies, accelerating the accumulation cycle. Customs officers have statistics that can be added to the questionnaires; tax officials, labour unions, geographers, journalists all produce a huge quantity of records, polls and charts. Those who sit inside the many Bureaus of Statistics may combine, shuffle around, superimpose and recalculate these figures and end up with a 'gross national product' or a 'balance of payments', exactly as others, in different offices, end up with 'Sakhalin island', 'the taxonomy of mammals', 'proven oil reserves' or 'a new planetary system'.

All these objects occupy the beginning and the end of a similar accumulation cycle; no matter whether they are far or near, infinitely big or small, infinitely old or young, they all end up at such scale that a few men or women can dominate them by sight; at one point or another, they all take the shape of a flat surface of paper that can be archived, pinned on a wall and combined with others; they all help to reverse the balance of forces between those who master and those who are mastered.

To be sure, expeditions, collections, probes, observatories and enquiries are only some of the many ways that allow a centre to act at a distance. Myriads of others appear as soon as we follow scientists in action, but they all obey the same selective pressure. Everything that might enhance either the mobility, or the stability, or the combinability of the elements will be welcomed and selected if it accelerates the accumulation cycle: a new printing press that increases the mobility and the reliable copying of texts: a new way to engrave by aquaforte more accurate plates inside scientific texts, a new projection system that allows maps to be drawn with less deformation of shape, a new chemical taxonomy that permits Lavoisier to write down the combinations of more elements, but also new bottles to chloroform animal specimens, new dyes to colour microbes in cultures, new classification schemes in libraries to find documents faster, new computers to enhance the weak signals of the telescopes, sharper styluses to record more parameters on the same electrocardiograms.[6] If inventions are made that transform numbers, images and texts from all over the world into the same binary code inside computers, then indeed the handling, the combination, the mobility, the conservation and the display of the traces will all be fantastically facilitated. When you hear someone say that he or she 'masters' a question better, meaning that his or her *mind* has enlarged, look first for inventions bearing on the mobility, immutability or versatility of the traces; and it is only later, if by some extraordinary chance, something is still unaccounted for, that you may turn towards the mind. (At the end of Part B, I will make this a rule of method, once a crucial element has been added.)

(3) Constructing space and time

The cumulative character of science is what strikes observers so much; why they devised the notion of a Great Divide between our scientific cultures and all the others. Compared to cartography, zoology, astronomy and economics, it seems that each ethnogeography, ethnozoology, ethnoastronomy, ethnoeconomics is peculiar to one place and strangely non-cumulative, as if it remained for ever stuck in a tiny corner of space and time. However, once the accumulation cycle and the mobilisation of the world it triggers are considered, the superiority of some centres over what appear by contrast to be the periphery may be documented without any additional divide between cultures, minds or logics. Most of the difficulties we have in understanding science and technology proceeds from our belief that space and time exist independently as an unshakable frame of reference *inside which* events and place would occur. This belief makes it impossible to understand how different spaces and different times may be produced *inside the networks* built to mobilise, cumulate and recombine the world.

For instance, if we imagine that the knowledge of Sakhalin island possessed by the Chinese fishermen is *included* in the scientific cartography elaborated by Lapérouse, then indeed it appears, by comparison, local, implicit, uncertain and weak. But it is no more included in it than the opinions about the weather are a sub-set of meteorology (see Chapter 5, Part A). Cartography is one network cumulating traces in a few centres which by themselves are as local as each of the points Lapérouse, Cook or Magellan cross; the only difference is in the slow construction of a map inside these centres, a map that defines two-way movement to and from the periphery. In other words, we do not have to oppose the local knowledge of the Chinese to the universal knowledge of the European, but only two local knowledges, one of them having the shape of a network transporting back and forth immutable mobiles to act at a distance. As I said in the Prologue, who includes and who is included, who localises and who is localised is not a cognitive or a cultural difference, but the result of a constant fight: Lapérouse was able to put Sakhalin on a map, but the South Pacific cannibals that stopped his travel put him on *their* map!

The same divide seems to take place between local ethnotaxonomy and 'universal' taxonomies as long as the networks of accumulation are put out of the picture. Can botany, for instance, displace all the ethnobotanies and swallow them as so many sub-sets? Can botany be constructed everywhere in a universal and abstract space? Certainly not, because it needs thousands of carefully protected cases of dried, gathered, labelled plants; it also needs major institutions like Kew Gardens or the Jardin des Plantes where living specimens are germinated, cultivated and protected against cross-fertilisation. Most ethnobotanies require familiarity with a few hundred and sometimes a few thousand types (which is already more than most of us can handle); but inside Kew Gardens, the new familiarity constituted by many sheets of neighbouring herbaries brought from all around the world by expeditions of all the nations of Europe requires the handling of tens and sometimes hundreds of thousands of types (which is too much for anyone to handle). So new inscriptions and labelling procedures have to be devised to limit this number again (see Part B). Botany is the *local knowledge* generated inside gathering institutions like the Jardin des Plantes or Kew Gardens. It does not extend further than that (or if it does, as we will see in Part C, it is by extending the networks as well).[7]

To go on in our journey we should force these immense extents of space and time generated by geology, astronomy, microscopy, etc., back inside their networks – these phentograms, billions of electrovolts, absolute zeros and eons of times; no matter how infinitely big, long or small they are, these scales are never much bigger than the few metre squares of a geological or an astronomical map, and never much more difficult to read than a watch. We, the readers, do not live *inside* space, that has billions of galaxies in it; on the contrary, this space is generated *inside* the observatory by having, for instance, a computer count little dots on a photographic plate. To suppose, for example, that it is possible to draw together in a synthesis the times of astronomy, geology, biology, primatology and anthropology has about as much meaning as making a synthesis between the pipes or cables of water, gas, electricity, telephone and television.

You are ashamed of not grasping what it is to speak of millions of light years?

Don't be ashamed, because the firm grasp the astronomer has over it comes from a small *ruler* he firmly applies to a *map* of the sky like you do to your road map when you go out for a camping trip. Astronomy is the local knowledge produced inside these centres that gather photographs, spectra, radio signals, infrared pictures, everything that makes a trace that other people can easily dominate. You feel bad because the nanometres of living cells baffle your mind? But it means nothing for anyone as long as it baffles the mind. It begins to mean something when the *nanometres* are *centimetres* long on the scaled-up electron photograph of the cell, that is when the eye sees it at the familiar scale and distance. Nothing is unfamiliar, infinite, gigantic or far away in these centres that cumulate traces; quite the opposite, they cumulate so many traces so that everything can become familiar, finite, nearby and handy.

It seems strange at first to claim that space and time may be constructed locally, but these are the most common of all constructions. Space is constituted by reversible and time by irreversible displacements. Since everything depends on having elements displaced each invention of a new immutable mobile is going to trace a different space-time.

When the French physiologist Marey invented at the end of the nineteenth century the photographic gun with which one could capture the movement of a man and transform it into a beautiful visual display, he completely reshuffled this part of space-time. Physiologists had never before been able to dominate the movement of running men, galloping horses and flying birds, only dead corpses or animals in chains. The new inscription device brought the living objects to their desks with one crucial change: the irreversible flow of time was now synoptically *presented* to their eyes. It had in effect become a space on which, once again, rulers, geometry and elementary mathematics could be applied. Each of Marey's similar inventions launched physiology into a new cumulative curve.

To take up an earlier example, as long as the Portuguese carracks disappeared en route, no space beyond the Bojador Cape could be pictured. As soon as they started to reversibly come and go, an ever-increasing space was traced around Lisbon. And so was a new time: nothing before could easily discriminate one year from another in this quiet little city, at the other end of Europe; 'nothing happened' in it, as if time was frozen there. But when the carracks started to come back with their trophies, booty, gold and spices, indeed things 'happened' in Lisbon, transforming the little provincial city into the capital of an empire larger than the Roman Empire. The same construction of a new history was also felt all along the coasts of Africa, India and the Moluccas; nothing would be the same again now that a new cumulative network brought the spices to Lisbon instead of Cairo. The only way to limit this construction of a new space-time would be to interrupt the movement of the carracks, that is, to build another network with a different orientation.

Let us consider another example of this construction, one that is less grandiose than the Portuguese expansion. When Professor Bijker and his colleagues enters the Delft Hydraulics Laboratory in Holland they are preoccupied by the shape that a new dam to be built in Rotterdam harbour – the biggest port in the world – should take. Their problem is to balance the fresh water of the rivers and sea water. So many dams have limited the outflow of the rivers that salt, dangerous for the precious floral culture, is penetrating further inland. Is the new dam going to affect the salt or the fresh water? How can this be known beforehand? Professor Bijker's answer to this question is a radical one. The engineers build a dam, measure the inflow of salt and fresh water for a few years for different weather and tide conditions; then they destroy the dam and build another one, start the measurements again, and so on, a dozen times until they have limited to the best of their ability the intake of sea water. Twenty years and many million florins later, the Hydraulics Lab is able to tell the Port Authority of Rotterdam with a high degree of reliability what shape the dam should have. Are the officials really going to wait twenty years? Are they going to spend millions of florins building and destroying wharfs, thus blocking the traffic of the busy harbour?

They do not need to, because the years, the rivers, the amount of florins, the wharfs, and the tides have been *scaled down* in a huge garage that Professor Bijker, like a modern Gulliver, can cross in a few strides. The Hydraulics Laboratory has found ways to render the harbour mobile, ignoring those features deemed irrelevant, like the houses and the people, and establishing stable two-way connections between some elements of the *scale model* and those of the full-scale port, like the width of the channel, the strength of the flows, the duration of the tides. Other features which cannot be scaled down, like water itself or sand, have been simply transferred from the sea and the rivers to the plaster basins. Every two metres captors and sensors have been set up, which are all hooked up on a big mainframe computer that writes down on millimetred paper the amount of salt and fresh water in every part of the Lilliputian harbour. Two-way connections are established between these sensors and the much fewer, bigger and costlier ones that have been put into the full-scale harbour. Since the scale model is still too big to be taken in at a glance, video cameras have been installed that allow one control room to check if the tide patterns, the wave-making machine and the various sluices are working correctly. Then, the giant Professor Bijker takes a metre-long plaster model of the new dam, fixes it into place and launches a first round of tides shortened to twelve minutes; then he takes it out, tries another one and continues.

Sure enough, another 'Copernican revolution' has taken place. There are not that many ways to master a situation. Either you dominate it physically; or you draw on your side a great many allies; or else, you try to be there before anybody else. How can this be done? Simply by reversing the flow of time. Professor Bijker and his colleagues *dominate* the problem, *master* it more easily than the port officials who are out there in the rain and are much smaller than the landscape. Whatever may happen in the full-scale space-time, the engineers will have *already seen it*. They will have become slowly acquainted with all the possibilities, rehearsing each scenario at leisure, capitalising on paper possible outcomes, which gives them years of experience more than the others. The order of time and space has been completely reshuffled. Do they talk with more authority and more certainty than the workmen building the real dam there? Well, of course, since they have already made all possible blunders and mistakes, safely inside the wooden hall in Delft, consuming only plaster and a few salaries along the way, inadvertently flooding not millions of hard-working Dutch but dozens of metres of concrete floor. No matter how striking it is, the superiority gained by Professor Bijker over the officials, architects and masons about the shape of the dam is no more supernatural than that of Marey, of the Portuguese or of the astronomer. It simply depends on the possibility of building a different space-time.

We now have a much clearer idea of what it is to follow scientists and engineers in action. We know that they do not extend 'everywhere' as if there existed a Great Divide between the universal knowledge of the Westerners and the local knowledge of everyone else, but instead that they travel inside narrow and fragile networks, resembling the galleries termites build to link their nests to their feeding sites. Inside these networks, they make traces of all sorts circulate better by increasing their mobility, their speed, their reliability, their ability to combine with one another. We also know that these networks are not built with homogeneous material but, on the contrary, necessitate the weaving together of a multitude of different elements which renders the question of whether they are 'scientific' or 'technical' or 'economic' or 'political' or 'managerial' meaningless. Finally, we know that the results of building, extending and keeping up these networks is to act at a distance, that is to do things in the centres that sometimes make it possible to dominate spatially as well as chronologically the periphery. Now that we have sketched the general ability of these networks to act at a distance and portrayed the mobilisation and accumulation of traces, there are two more problems to tackle: what is done *in* the centres and *on* the accumulated traces that gives a definitive edge to those who reside there (Part B); and what is to be done to maintain the networks in existence, so that the advantages gained in the centres have some bearing on what happens at a distance (Part C).

Part B
Centres of calculation

After having followed expeditions, collections and enquiries, and observed the setting up of new observatories, of new inscription devices and of new probes, we are now led back to the centres where these cycles started from; inside these centres, specimens, maps, diagrams, logs, questionnaires and paper forms of all sorts are accumulated and are used by scientists and engineers to escalate the proof race; every domain enters the 'sure path of a science' when its spokespersons have so many allies on their side. The tiny number of scientists is more than balanced by the large number of resources they are able to muster. Geologists can now mobilise on their behalf not a few rocks and a few nice water colours of exotic landscapes, but hundreds of square metres of geological maps of different parts of the earth. A molecular biologist, when she talks of mutations in maize, may now have at her side not a few wild cobs, but protocol books full of thousands of cross-breeding results. The directors of the Census Bureau now have on their desks not only newspaper clippings with opinions on how big and rich their country is, but stocks of statistics extracted from every village that array their countrypeople by age, sex, race and wealth. As for astronomers, a chain of radio-telescopes working together transforms the whole earth into one single antenna that delivers thousands of radio sources through computerised catalogues to their offices. Every time an instrument is hooked up to something, masses of inscriptions pour in, tipping the *scale* once again by forcing the world to come to the centres – at least on paper. This mobilisation of everything that can possibly be inscribed and moved back and forth is the staple of technoscience and should be kept in mind if we want to understand what is going on inside the centres.

(1) Tying all the allies firmly together

When entering the many places where stable and mobile traces are gathered, the first problem we will encounter is how to *get rid of them*. This is not a paradox, but simply an outcome of the setting up of instruments. Each voyage of exploration, each expedition, each new printer, each night of observation of the sky, each new poll, is going to contribute to the generation of thousands of crates of specimens or of sheets of paper. Remember that the few men and women sitting inside Natural History Museums, Geological Surveys, Census Bureaus or other laboratories do not have especially huge brains. As soon as the number or the scale of elements to handle increases, they get lost like anyone else. The very success of the mobilisation, the very quality of the instruments, will have as its first consequence their drowning in a flood of inscriptions and specimens. By itself, the mobilisation of resources is no guarantee of victory; on the contrary, a geologist surrounded by hundreds of crates full of unlabelled fossils is in no better position to dominate the earth than when he was in Patagonia or in Chile. This flooding of investigators by the inscriptions is, so to speak, a revenge of the mobilised world. 'Let the earth come to me, instead of me going to the earth,' says the geologist who starts a Copernican revolution. 'Very well,' answers the earth, 'here I am!' The result is utter confusion in the basement of the building of the Geological Survey.

Because of this situation, *additional* work has to be done inside the centres to mop up the inscriptions and reverse the balance of forces once more. I defined above the stability of the traces as the possibility of going back and forth from the centres to the periphery; this feature is all the more essential when going from primary traces to second degree traces that make possible the handling of the first.

(A) SOLVING A FEW LOGISTICAL PROBLEMS

For instance, the director of the census cannot be confronted at the same time
with the 100 million questionnaires brought in by the pollsters. He would see
nothing but reams of paper – and, to begin with, he will be unable to know how
many questionnaires there are. One solution is to do to the questionnaires what
the questionnaires did to the people, that is, to extract from them some elements
and to place them on another more mobile, more combinable paper form. This
operation of ticking rows and columns with a pencil is a humble but a crucial one;
in effect, it is the same operation through which what people said to the pollster
was transformed into boxes of the questionnaire or through which Sakhalin island
was transformed by Lapérouse into latitude and longitude on a map.

In all cases the same problem is partially solved: how to keep your informants
by your side while they are far away. You cannot bring the people to the Census
Office, but you can bring the questionnaires; you cannot display all the
questionnaires, but you can show a tally where each answer to the questionnaire
is represented by a tick in a column for sex, age, etc. Now, a new problem will
emerge if the tallies are carefully done: you will obtain too many marks on too
many columns for even the best mind to embrace them all at once. Thus you will
be swamped again in paper forms exactly as you were with the questionnaires and
earlier with the people. A third degree paper form is now necessary to record not
the marks any more, but the *totals* at the bottom of each row and column.
Numbers are one of the many ways to sum up, to summarise, to totalise – as the
name 'total' indicates – to bring together elements which are, nevertheless, not
there. The phrase '1,456,239 babies' is no more made of crying babies than the
word 'dog' is a barking dog. Nevertheless, once tallied in the census, the phrase
establishes *some* relations between the demographers' office and the crying
babies of the land.

However, the flood is going to be shifted somewhere else in the Census Bureau,
because too many totals are now pouring in from the thousands of marks in
columns or from holes in punch cards. New fourth degree inscriptions
(percentages for instance or graphs or pie charts) have to be devised to mop up
the totals again, to mobilise them in a displayable form whilst still retaining some
of their features. This cascade of fourth, fifth and *nth* order inscription will never
stop, especially if the population, the computers, the profession of demography,
statistics and economics, and the Census Bureau all grow together. In all cases,
the *nth* order inscription will now *stand for* the *nth* -1 order paper forms exactly as
these in turn *stood for* the level just below. We know from earlier chapters that
these translations and representations may be disputed, but this is not the point
here; the point is that, in case of a dispute, other tallies, code words, indicators,
metres and counters will allow dissenters to go back from the *nth* final inscription
to the questionnaires kept in the archives and, from it, to the people in the land.
That is, some two-way relations have been established between the desk of the
director and the people, relations that allow the director, if there is no dissenter,

to engage in some controversies as if speaking in the name of his millions of well-arrayed and nicely displayed allies.

This example is enough to define the additional work necessary to transform the inscriptions. What shall we call this work? We could say that the task is to make the many act as one; or to establish longer networks; or to simplify yet again the inscriptions; or to build up a cascade of successive representatives; or to 'punctualise' a multitude of traces; or to simultaneously mobilise elements while keeping them at a distance. Whatever we call it, the general shape is easy to grasp: people inside the centres are busy building elements with such properties that when you hold the final elements you also, in some way, hold the others, building, in effect, *centres inside the centres*.

One more example will give a more precise idea of this additional work, which should not be severed from the rest of network building. When they organised their first international meeting in Karlsruhe in 1860, European chemists were in a state of confusion similar to the one I sketched above, because every new school of chemistry, every new instrument was producing new chemical elements and hundreds of new chemical reactions.[8] Lavoisier listed thirty-three simple substances, but with the introduction of electrolysis and spectral analysis, the list has increased to seventy at the time of the meeting. To be sure the cascade of transformations was already well under way; each substance had been renamed and labelled with a common tally (its atomic weight, standardised at the Karlsruhe meeting), allowing chemists to write down lists of substance and to rank them in multiple ways, but this was not enough to dominate the multiplicity of reactions. As a result, introductory courses to the newly professionalised chemistry were made of long and rather chaotic lists of reactions. To remedy this confusion, dozens of chemists were at the time busy classifying chemical substances, that is drawing some sort of tables with columns so devised that, considering them synoptically, chemistry could be embraced, in the same way as the earth can be overviewed on a map or a nation through statistics. Mendeleev, who had been asked to write a chemistry textbook, was one of them. Believing that it was possible to find a real classification and not to write down a mere stamp collection he distinguished 'substance' from 'element'. He wrote each element down on a card, and shuffled the pack as in a patience game, trying to find some recurring pattern.

There is no reason to give up following scientists simply because they are handling paper and pencil instead of working in laboratories or travelling through the world. The construction of *nth* order paper form is no different from the *nth −1* – although it is sometimes more elusive and much less studied. The difficulty of this new patience game invented by Mendeleev is not only to look for a pattern by lines and columns that would include all the elements – everyone else had done this before; the difficulty is to decide, in cases where some element does not fit, or when there is no element to fill in a box, if the drafted table is to be discarded or if the missing elements are to be either brought from elsewhere or discovered later. After long struggles between different tables and many counter-examples, Mendeleev settled, on March 1869, on a compromise that satisfied him: a table that listed elements by their atomic weight and ranked them vertically by their valences, requiring only several elements to be displaced and several others to be found. Each element is now situated on a new paper form at the intersection of a longitude and of a latitude; those on the same horizontal line are close by their atomic weight although foreign by their chemical properties; those on the same vertical line are similar by their properties although they are more and more distant by their atomic weight. A new space is thus locally created; new relations of distance and proximity, new neighbourhoods, new families are devised: a periodicity (hence the name of his table) appears which was invisible until then in the chaos of chemistry.

At each translation of traces onto a new one something is *gained*. Louis XVI at Versailles can *do* things with the map (for instance draw boundaries to partition the Pacific) that neither the Chinese nor Lapérouse could do; Professor Bijker can become familiar with the future of Rotterdam harbour (for instance checking its resistance to an elevation of the North Sea) *before* the officials, the sailors and the North Sea; demographers can *see* things on the final curve summarising the census (for instance age pyramids) that none of the pollsters, none of the politicians, none of the interviewed people could see before; Mendeleev can gain in advance some *familiarity* with an empty box of his table before the very people who discovered the missing elements (like Lecoq de Boisbaudran with gallium occupying the box left vacant in the table under the name of eka-aluminum).[9]

It is important for us to do justice to the cleverness of this additional work going on in the centres without exaggerating it and without forgetting that it is just that: *additional* work, a slight enhancement of one of the three qualities of the inscriptions, namely mobility, stability and combinability. First, the gain does not always offset the losses that are entailed by the translation of one form into another (see Part C): holding the map in Versailles did not protect Louis XVI's possessions from being taken over by the English; there is no guarantee that the events of the Delft scale model will be mimicked by the Rotterdam harbour in the next century; planning an increasing birth rate in the Census Bureau is not exactly like conceiving new babies; as to Mendeleev's table it was to be soon disrupted by the emergence of radioactive chemical monsters he could not place. Second, when there is a gain, it is not supernatural power brought to the scientists by an angel sent straight from Heaven. The gain is *on* the paper form itself. For instance, the supplement offered by the map is *on* the flat surface of paper which is easily dominated by the eyes and on which many different elements can be painted, drawn, superimposed and inscribed. It was calculated that drawing a map of England with 200 towns (that is an input of 400 longitudes and latitudes) allows you to trace 20,000 itineraries from one town to another (thus yielding an output 50 to 1!).[10] Similarly, the empty boxes in Mendeleev's table are offered to him *by* the geometrical pattern of rows and columns. To be sure, his success in anticipating unknown elements to fill in the boxes is an impressive one. What is also extraordinary is how chemical reactions taking place in gallipots and stills all over Europe have been brought to bear on a simple pattern of rows and columns through a long cascade of translations. In other words, the *logistics* of immutable mobiles is what we have to admire and study, not the seemingly miraculous supplement of force gained by scientists thinking hard in their offices.

(B) CALCULATING, AT LAST . . .

Inside the centres, logistics requires the fast mobilisation of the largest number of elements and their greatest possible fusion. Tallies, totals, graphs, tables, lists, these are some of the tools that make the additional treatment of inscriptions possible. There exist a few others that have received both too much and too little attention. Too much because they are the object of a cult; too little because too few people have studied them dispassionately. As a consequence there is not a large body of empirical literature on which we can rely to guide our travel, as we could in the other chapters. When we reach the realm of calculations and theories we are left almost empty-handed. In the remainder of this Part, I must confess that what is left is a programme of research, not an accumulation of results; what is left is obstinacy, not resources.

The risk of the cascade I presented above is of ending up with a few manageable but meaningless numbers, insufficient at any rate in case of controversy since, in effect, the allies have deserted in the meantime. Instead of a capitalisation, the centres would end up with a net loss. The ideal would be to retain as many elements as possible and still be able to manage them. Statistics is a nice example of those devices that simultaneously solve the two problems. For instance, if I give the director of the census the *mean* population increase in the land, he will be interested but also disappointed because he lost in the process the dispersion (the same mean could be obtained by a few eight-child families or by a lot of two-and-a-half-child families). The simplification has been such that the director gets only an impoverished version of the census. If a new calculation is invented that keeps, through the various simplifications, both the mean and the dispersion of the data, then part of the problem is solved. The invention of *variance* is one of these devices that continue to solve the major problems of the inscriptions: mobility, combinability and faithfulness. So is the invention of sampling. What is the minimum sample that allows me to represent the largest number of features? Statistics, as its name and history indicates, is the science of spokespersons and statesmen *par excellence*.[11]

Let us take as another example, the work of Reynolds, a British engineer specialising in fluid mechanics who, at the turn of the century, studied the complex problem of turbulence.[12] How can you *relate* the many instances of turbulence observed in scale models, or along rivers? These instances are already summed up in sentences of the form 'the more . . . the more', 'the more . . . the less'. The faster the flow the more turbulence there is; the bigger the obstacle encountered by the flow the more turbulence there will be; the denser a fluid the more prone to turbulence it is; finally, the more viscous a fluid the less turbulence there will be (oil turns smoothly around an obstacle that would have triggered eddies in water). Can these sentences be more firmly tied together in an *n+1* inscription? Instead of ticking boxes in tables we are going to give a letter to each of the relevant words above and replace the comparatives 'more' and 'less' by multiplication and division. The new summary has now this shape:

T(urbulence) is proportional to S(peed)
T is proportional to L(ength of the obstacle)
T is proportional to D(ensity)
T is inversely proportional to V(iscosity), or $T \frac{1}{V}$

This new translation does not seem to add much; except that it can now be displayed synoptically in a still shorter form:

$$T \text{ (is related to) } \frac{SLD}{V}$$

There is no great gain yet; the new summary simply states that there exist tight relations between these elements and indicates roughly what type of relation it is. After some fiddling around so that the units compensate one another and give a non-dimensional number, Reynolds ends up with a new formula:

$$R = \frac{SLD}{V}$$

Is there anything gained by holding Reynolds's formula or is it simply an abridged summary of all the instances? As in Mendeleev's table, and indeed all the rewritings observed in this section, something is gained because each translation reshuffles the connections between elements (thus creating a new space-time). Situations which appeared as far apart as a fast little creek running against a stone and a big slow river stopped by a dam, or a feather falling in the air and a body swimming in molasses, may produce turbulences which look the same if they have the 'same Reynolds' (as it is now called). *R* is now a *coefficient* that can label all possible turbulences, whether galaxies in the sky or knots on a tree, and it does indeed, as the name 'coefficient' reminds us, make all turbulences act as one in the physicist's lab. Better still, the Reynolds number allows Professor Bijker in his laboratory, or an aircraft engineer in a wind tunnel, to decide how to scale down a given situation. As long as the scale model retains the same Reynolds as the full-scale situation, we can work on the model even if it 'looks' entirely different. Differences and similarities are recombined as well as what types of inscriptions one should believe *more* than others.

Although this is indeed a decisive advantage provided by what is aptly called an *equation* (because it ties different things together and makes them equivalent), this advantage should not be exaggerated. First, an equation is no different in nature from all the other tools that allow elements to be brought together, mobilised, arrayed and displayed; no different from a table, a questionnaire, a list, a graph, a collection; it is simply, as the end-point of a long cascade, a means to accelerate the mobility of the traces still further; in effect, equations are *sub-*

sets of translations and should be studied like all the other translations. Second, it cannot be severed from all the network-building, of which it is but a tiny part; for instance, the Reynolds number allows scientists to go from one scale model to another and to travel fast from one instance of turbulence to another far away in space and time; very well, but it works only as long as there are hundreds of hydraulic engineers working on turbulences (and they, in turn, work on scale models only in so far as their laboratories have been able to become involved with the construction of harbour, dams, pipes, aircrafts, etc.). It is only once the networks are in place that the invention of the Reynolds number can make a difference. To make a metaphor, it plays the same role as a *turntable* in the old railway system; it is important but you cannot reduce the whole system to it, since it takes such an important role *because and as long as* the mobilisation is under way (turntables, for instance, became irrelevant once electric traction allowed engines to go both directions).

Equations are not only good at increasing the mobility of the capitalised traces, they are also good at enhancing their combinability, transforming centres into what I will call centres of calculation. Such a centre was built by Edison at Menlo Park where the famous incandescent lamp at the end of the 1870s was invented.[13] Thanks to Edison's notebooks it is possible not only to reconstruct his strategy, not only to follow how his laboratory was constructed, but also to observe his work with paper and pencil on the *nth* degree inscriptions. No more than in the story of King John (see Part A) or in any other case should the 'intellectual' work be severed from the network-building in which Edison is engaged. His strategy is to substitute his company for gas companies, which means elaborating a complete system to produce and deliver electricity everywhere at the same consumer cost as that of gas. As early as 1878, Edison started work on the most classical of calculations: accounting and basic economics. How much would his projected system cost given the price of steam engines, dynamos, engineers, insurance, copper and so on? One result of his first paper estimate showed that the most expensive item was that of the copper necessary for the conductors. The price of copper was so high that, from the start, it made electricity unable to compete with gas. Thus something had to be done with copper.

Now comes the main logistical advantage provided by writing down all inscriptions in equation form. To calculate how much copper he needed, Edison not only used accounting but also one of Joule's equations (an equation obtained earlier through a process similar to the one I sketched with Reynolds): energy loss is equal to the square of the current multiplied by the length of the conductor multiplied by a constant, all divided by the cross-section of the conductor.

What is the relation between physics and economics? None if you consider Joule's laboratory on the one hand and physical plants on the other. In Edison's notebook, however, they progressively merge in one seamless cloth because they are all written in more or less the same form and presented synoptically to his eyes. The web of associations on which Edison works is *drawn together* by the equations. By manipulating the equations, he retrieves sentences like: the more you increase the cross-section to reduce loss in distribution, the more copper you will need. Is this physics, economics or technology? It does not matter, it is one single web that translates the question 'how do you bring down the price of copper' into 'how can you fiddle with classic equations of physics'. Edison is now surrounded by a set of heterogenous constraints; he tries to find out which is stronger and which weaker (see Chapter 5). The consumer price has to be equal to that of gas, this is an absolute requirement; so is the present price of copper on the market; so is Joule's law; so is Ohm's law that defines resistance as voltage divided by current

$$\text{Resistance} = \frac{\text{Voltage}}{\text{Current}}$$

Of course, if the current could be decreased, the cross-section could be decreased as well, and so will the copper bill. But according to Ohm's law this would mean *increasing* the resistance of the filament, that is looking for a *high*-resistance lamp when everyone at the time was looking for a *low*-resistance lamp because of the difficulty of finding a filament that would not burn out. Is this constraint as absolute as the others? Edison now tries out this chain of associations and evaluates how absolute it is. The equation above does not escape from the network in which Edison is placed; it is not because it is written in mathematical terms that we are suddenly led into another world. Quite the contrary, it *concentrates* at one point what the network is made of, what are its strong and weak points. Compared to the others, the amount of resistance appears to be the weakest link. It has to give way. No matter how difficult it appears, Edison decides, we will look for a high resistance lamp *because* it is the only way to maintain all the other elements in place. Once the decision is taken, Edison sends his troops on a trial-and-error one-year-long search for a filament that resists without burning out. The incandescent high-resistance lamp is the final result of the calculation above.

This example shows not only how foreign domains can be combined and brought to bear on one another once they have the common form of calculation. It also reveals the final and main advantage of equations. From the beginning of this book I have constantly presented scientists and engineers as mobilising large numbers of allies, evaluating their relative strength, reversing the balance of forces, trying out weak and strong associations, tying together facts and mechanisms. In effect, I had to replace each traditional divide by a relative distinction between stronger and weaker associations. We have now come close to the end of our long journey because the equations produced at the final edge of the capitalisation constitute, literally, the *sum* of all these mobilisations, evaluations, tests and ties. They tell us what is associated with what; they define the nature of the relation; finally, they often express a measure of the resistance of each association to disruption. Of course, they are utterly impossible to understand without the mobilisation process (and this is why I did not talk of them earlier), they are nevertheless the true heart of the scientific networks, more important to observe, study and interpret than facts or mechanisms, because they *draw* all of them together inside the centres of calculation.

(2) *What's the matter of (with) formalism?*

Following the cascade of inscriptions drawn by scientists, we have reached a point which should be the easiest of our trip since we can now reap the benefits of our earlier work on weaker and stronger associations. Unfortunately, it is also the part which has been somewhat obscured by earlier investigators, which means that we still have to be very careful in defining what we have to study and whom we have to follow. Two confusing words have been used before to account for what happens in the centres of calculation: those of abstraction and theory. Let us examine what they mean.

(A) DOING AWAY WITH 'ABSTRACT THEORIES'

In the cascades that we followed in the section above, we always went from one practical and localised activity to another; to be sure, each stage of translation simplified, punctualised and summarised the stage immediately below. But this activity of re-representation[14] of the supporters was very concrete indeed; it required pieces of paper, laboratories, instruments, tallies, tables, equations; above all, it was imposed by the necessity of mobilisation and action at a distance, and never abandoned the narrow networks that made it possible. If by 'abstraction' is meant the process by which each stage extracts elements out of the stage below so as to gather in one place as many resources as possible, very well, we have studied (and continue to study) the process of abstraction, exactly as we would examine a refinery in which raw oil is cracked into purer and purer oils. Alas, the meaning of the word 'abstraction' has shifted from the *product* (*nth* order inscriptions) to not only the *process* but also to the *producer's mind*. It is thus implied that scientists in the centres of calculations would think 'abstractly', or at least more abstractly than others. Lapérouse will be said to operate more abstractly than the Chinese when he handles latitudes and longitudes, and Mendeleev to think more abstractly than an empirical chemist when he shuffles his cards around. Although this expression has as much meaning as saying that a oil refinery refines petrol 'refiningly', it is enough to fog the issue. The concrete work of making abstractions is fully studiable; however, if it becomes some mysterious feature going on in the mind then forget it, no one will ever have access to it. This confusion between the refined product and the concrete refining work is easy to clarify by using the substantive 'abstraction' and never the adjective or the adverb.

However, this simple rule of hygiene is made harder to apply because of the cult of 'theories'. If by 'theory' is meant the crossroads that allow the centres to mobilise, manipulate, combine, rewrite and tie together all the traces obtained through the ever-extending networks, then we should be able to study theories fully. As I said, they are centres inside the centres providing one more acceleration of the mobility and combinability of the inscriptions. Studying them should be no more difficult than understanding the role of clover-leaf intersections when examining the American freeway system, or the function of digital telephone exchanges when observing the Bell network. If the mobilisation increases in scale, then, necessarily, the products at the intersection of all networks have to be enhanced. Any innovation at these intersections will give a decisive edge to the centres.

This situation is altered if the meaning of the word 'theory' shifts to become an adjective or an adverb (some people are then said to handle more 'theoretical' matters or to think 'theoretically'), but it is much worse when 'theories' are transformed into 'abstract' objects severed from the elements they tie together. This happens for instance if Mercator's work in finding a new geometrical projection for navigational maps is disconnected from the navigators' travels; or if Mendeleev's table is cut off from the many of chemists' elements he tried to tie together in one coherent whole; or if Reynold's number is cut off from the experimental turbulences that he was trying to classify with one single coefficient. As soon as a divide is made between theories and what they are theories *of*, the tip of technoscience is immediately shrouded in fog. Theories, now made abstract and autonomous objects, float like flying saucers above the rest of science, which by contrast becomes 'experimental' or 'empirical'.[15]

The worst is yet to come. Since sometimes it happens that these abstract theories, independent of any object, nevertheless have some bearing on what happens down below in empirical science – it has to be a *miracle*! Miracle indeed to see a clover-leaf intersection fitting *precisely* with the freeways whose flow it redistributes! It is amusing to see rationalists admire a miracle of that quality while they deride pilgrims, dervishes or creationists. They are so enthralled by this mystery that they are fond of saying, 'The least understandable thing in the world is that the world is understandable.' Speaking about theories and then gaping at their 'application' has no more sense than talking of clamps without ever saying what they fasten together, or separating the knots from the meshes of a net. Doing a history of scientific 'theories' would be as meaningless as doing a history of hammers without considering the nails, the planks, the houses, the carpenter and the people who are housed, or a history of cheques without the bank system. By itself, however, the belief in theory would not impress much if it were not reinforced by the trials in responsibility we learned to study in Chapters 3 and 4. As the reader may recall, the result of these trials was to make the few scientists at the end of the mobilisation process responsible for the whole movement. When the two processes are compounded, we get not only the assertion that scientists lead the world but that scientists' theories lead the world! The pyramid of Cheops is now standing on its tip, which does make the world quite hard to understand.

A few common-sense precepts will be enough to put the pyramid back on its base. First, we will abstain from ever using the words 'abstraction' and 'theory' in adjectival or adverbial forms. Second, we will never cut off the abstractions or the theories from what they are abstractions or theories *of*, which means that we will always travel through the networks along their greatest length. Third, we will never study a calculation without studying the *centres* of calculation. (And, of course, as we learned earlier, we will not confuse the results of the attribution process with the list of those who actually did the job.)

(B) WHY FORMS MATTER SO MUCH: THE SEVENTH RULE OF METHOD

Perhaps it would be best to do away altogether with the tainted words 'abstraction' and 'theory'. However, even if it is easy to do away with them, and with the cult rendered to them, we still have to account for the phenomena they point at so clumsily.

As we saw in section 1, the construction of the centres requires elements to be brought in from far away – to allow centres to dominate at a distance – *without* bringing them in for good – to avoid centres being flooded. This paradox is resolved by devising inscriptions that retain simultaneously as little and as much as possible by increasing either their mobility, stability or combinability. This compromise between presence and absence is often called **information**. When you hold a piece of information you have the *form* of something without the thing itself (for instance the map of Sakhalin without Sakhalin, the periodic table without the chemical reactions, a model of Rotterdam harbour without the harbour itself). As we know, these pieces of information (or forms, or paper forms, or inscriptions – all these expressions designate the same movement and solve the same paradox) can be accumulated and combined in the centres. But their accumulation has one more unexpected *by-product*. Since there is no limit to the cascade of rewriting and re-representation, you may obtain *nth* order forms that are combined with other *nth* order forms coming from completely different regions. It is these new unexpected connections that explain why forms matter so much, and why observers of science are so thrilled with them.

First, we have to remove one little mystery: how is it that the 'abstract' forms of mathematics apply to the 'empirical world'? Many books have been written to find an explanation to this 'well-known fact' but almost no one has bothered to verify its existence. If the practice of science was followed, however, it would be quickly apparent that it never happens. 'Abstract' mathematics never applies to the 'empirical world'. What happens is much more clever, much less mystical and much more interesting. At a certain point in the cascade, instruments start to inscribe forms on, for example, graph paper. A cloud of points obtained from the census through many transformations ends up, after a few more statistical rearrangements, as a line on a graph. Interestingly enough, amino acid analysers also display their results on a graph paper. More curiously, Galileo's study of a falling body also takes the form of a graph (when it is repeated today) and had the shape of a triangle in his own notebooks.[16] Mathematics might be far from households, amino acids, and wooden spheres rolling over an inclined plane. Yes, but once households, amino acids and inclined planes have been, through the logistics above, brought onto a white piece of paper and asked to write themselves down in forms and figures, then their mathematics is very, very close; it is literally as close as one piece of paper is from another in a book. The adequation of mathematics with the empirical world is a deep mystery. The superimposition of one mathematical form on paper and of another mathematical form drawn on the printout of an instrument is not a deep mystery, but is quite an achievement all the same.[17]

Were we to follow how the instruments in the laboratories write down the Great Book of Nature in geometrical and mathematical forms we might be able to understand why forms take so much precedence. In the centres of calculation, you obtain paper forms from totally unrelated realms but with the same shape (the same Cartesian coordinates and the same functions, for instance). This means that *transversal* connections are going to be established in addition to all the *vertical* associations made by the cascade of rewriting. Thus, someone who wants to work on functions would be able to intervene, in a few years, in ballistics, demography, the revolution of planets, card games, in anything – as long as it has been first displayed in Cartesian coordinates.

The very growth of the centres entails the multiplication of instruments which, in turn, oblige the information to take a more and more mathematical shape on paper. This means that the calculators, whoever they are, sit at a central point inside the centres because everything has to pass through their hands.

For instance, once Sakhalin is put on the map, you can apply on a *flat* surface of paper a graduated ruler and a compass and calculate a possible route: 'If a ship arrives from this point, she will sight the land at 20° NNE after a route of 120 nautical miles keeping her course at 350°.' Or can you? Well, it all depends on how the package of bearings sent by Lapérouse is put on the map. Exactly as Lapérouse transformed the Chinese talk into a list of two-figure readings (longitudes and latitudes), this list is now transformed into points on a *curved* surface figuring the earth. But how to go from the curved to the flat surface without further deformation? How to maintain the information through all these transformations? This is a very concrete and practical problem, but neither Lapérouse nor his Chinese informants can solve it. This is the sort of question that can be solved only in the centres by people working on *nth* order forms, wherever they come from. The problem above is now translated into another one: how to project a sphere on a surface? Since something will be lost in the projection, what should I keep? The angles or the surface? Mercator's choice was to keep the angles so important for deciding on the ships' routes and to give up the accurate rendering of the surfaces, which is interesting only for landsmen. The point is that, once the network is in place that in some ways ties together Lapérouse's travels and the cartographer's office, the *smallest change* in the geometry of projection might have enormous consequences since the flow of forms coming from all over the planet and back to all the navigators will be altered. The tiny projection system is an obligatory passage point for the immense network of geography. Those who sit at this point, like Mercator, carry the day.

When people wonder how 'abstract' geometry or mathematics may have some bearing on 'reality', they are really admiring the *strategic position* taken by those who work inside the centres on forms of forms. They should be the weakest since they are the most remote (as it is often said) from any 'application'. On the contrary, they may become the strongest by the same token as the centres end up controlling space and time: they design networks that are tied together in a few obligatory passage points. Once every trace has been not only written on paper, but rewritten in geometrical form, and re-written in equation form, then it is no wonder that those who control geometry and mathematics will be able to intervene almost anywhere. The more 'abstract' their theory is, the better it will be able to occupy centres inside the centres. When Einstein is preoccupied by clocks and how to reconcile their readings when they are so far apart that it takes time for the observer of one clock to send the information to another observer, he is not in an abstract world, he is deep down at the centre of all exchanges of information, attentive to the most material aspect of inscription devices: How do I know what time it is? How do I see that there is superimposition of the hands of the clock? What should I give up if I wish to maintain above all the equivalence of all the observers' signals in case of great speed, great masses and great distance? If the centres of calculation wish to handle all the information all travellers on ships bring them, they need Mercator and his 'abstract' projection; but if they wish to handle systems that travel at the speed of light and still maintain the stability of their information, they need Einstein and his 'abstract' relativity. Giving up a classic representation of the space-time is not too high a price if the pay-off is a fantastic acceleration of the traces and an enhancement of their stability, faithfulness and combinability.

At the limit, if mathematicians stop talking of equations and geometry altogether, and start considering 'number' *per se*, 'set' in general, 'proximity', 'association', the more *central* their work will become since it will concentrate still further what is going on in the centres of calculation. The sheer accumulation of *nth* order paper forms makes any *nth+1* form that can at the same time maintain the features and get rid of the thing (of the 'matter') relevant. The more heterogeneous and dominating the centres, the more formalism they will require simply to stay together and maintain their imperium. Formalism and mathematics are attracted by the centres, if I dare make this metaphor, like rats and insects by granaries.

If we wish to follow scientists and engineers to the end, we will have to penetrate, at one point or another, what has become the Holy of Holies. Only a few features are clear at this point. First, we do not have to suppose a priori that formalism escapes from the mobilisation, from the centres, from the network-building. It is not transcendental, as philosophers say to account for the incredible

supplement of forces it provides to those who develop them. This supplement
gained from manipulating *nth* degree forms comes entirely from inside the
centres and is probably better accounted for by the many new transversal
connections it allows. Second, we do not have to lose our time finding empirical
counterparts to explain these forms by simple, practical manipulations, similar
to the ones done outside the centres. The handling of pebbles on Sakhalin beach
will never give you set theory or topology. To be sure, the cascade of inscriptions
is a practical and concrete manipulation of paper forms all along, but each end-
product is a form that does not resemble anything on the level below – if it does, it
means this rung in the ladder is useless, that at least that part of the translation
has failed. Third, we do not have to waste any time looking for 'social
explanations' of these forms, if by social is meant features of society mirrored by
mathematics in some distorted way. Forms do not distort or misrepresent
anything, they accelerate still more the movement of accumulation and
capitalisation. As I have hinted all along, the link between society and
mathematics is both much more distant and much more direct than expected:
they explicitly attach firmly together all possible allies, constituting in effect what
is probably the hardest and most 'social' part of society. Fourth, there is no
reason to fall back on conventions that scientists would agree with one another in
order to account for the bizarre existence of these forms that seem *unrelated* to
anything else. They are no less real, no more sterile, no more pliable than any
other inscriptions devised to make the world mobile and to carry it to the centres.
If anything, they resist more than anything else (by our definition of reality) since
they multiply and enhance the relations of all the other elements of the networks.
Fifth, to find our way, we have to take the grain of truth offered by each of these
four traditional interpretations of forms (transcendentalism, empiricism, social
determinism and conventionalism): *nth* order forms give an *unexpected*
supplement – as if coming from another world; they are the result of a *concrete*
work of purification – as if related to practical matters; they concentrate the
associations still more – as if they were *more social* than society; they tie together
more elements – as if they were *more real* than any other convention passed
among men.

Frankly, I have not found one single study which could fulfil this fifth
requirement. From this absence, one could draw the conclusion that forms
cannot be studied through any sort of enquiry like the one I have portrayed in this
book because they escape for ever what happens in the centres of calculation. But
I draw a different conclusion; almost no one has had the courage to do a careful
anthropological study of formalism. The reason for this lack of nerve is quite
simple: a priori, before the study has even started, it is towards the mind and its
cognitive abilities that one looks for an explanation of forms. Any study of
mathematics, calculations, theories and forms in general should do quite the
contrary: first look at how the observers move in space and time, how the
mobility, stability and combinability of inscriptions are enhanced, how the
networks are extended, how all the informations are tied together in a cascade of

re-representation, and if, by some extraordinary chance, there is something still unaccounted for, then, and only then, look for special cognitive abilities. What I propose, here, as a **seventh rule of method**, is in effect a *moratorium* on cognitive explanations of science and technology! I'd be tempted to propose a ten-year moratorium. If those who believe in miracles were so sure of their position, they would accept the challenge.

Part C
Metrologies

Translating the world towards the centres is one thing (Part A); gaining an unexpected supplement of strength by working inside these centres on *nth* degree inscriptions is another (Part B). There is still one remaining snag, because the final inscriptions are not the world: they are only representing it in its absence. New infinite spaces and times, gigantic black holes, minuscule electrons, enormous economies, mind-boggling billions of years, intricate scale models, complex equations, all occupy no more than a few square metres that a few per cent of the population (see Chapter 4) dominate. To be sure, many clever traps and tricks have been discovered to reverse the balance of forces and make the centres bigger and wiser than the things that dominated them until then. However, nothing is irreversibly gained at this point if there is no way to translate *back* the relation of strength that has been made favourable to the scientists' camp. More additional work has yet to be done. This movement from the centre to the periphery is to be studied as well, if we want to follow scientists up to the end. Although this last leg of the journey is as important as the other two, it is usually forgotten by the observers of science because of this queer notion that 'science and technology' are 'universal'; according to this notion, once theories and forms have been discovered, they spread 'everywhere' without added cost. This application of abstract theories everywhere and at every time appears to be another miracle. As usual, following scientists and engineers at work gives a more mundane but more interesting answer.

(1) Extending the networks still further

When, on 5 May 1961, Alan Shepard got his turn on the first American Mercury space flight, was it the *first time*?[18] In a way, yes, since no American had really been out there. In another sense, no, it was simply the *(n + 1)th* time. He had done every possible gesture hundreds of times before on the *simulator*, a scale model of another sort. What was his main impression when he finally got outside the simulator and inside the rocket? It was either 'just the way it sounded in the centrifuge' or 'it was different from the simulator, it was easier' or 'Man, that wasn't like the centrifuge, it was more sudden'. During his short flight he kept comparing the similarities and slight differences between the *nth* rehearsal on the flight simulator, and the *(n + 1)th* actual flight. The attendants in the control tower were surprised how cool Shepard was. This guy obviously had the 'right stuff' since he was not afraid of going out there in the unknown. But the point is that he was not really going into the unknown, as Magellan did crossing the strait that bears his name. He had been *there* already hundreds of times, and monkeys before him hundreds of other times. What is admirable is not how one can get into space, but how the complete space flight can be simulated in advance, and then slowly extended to unmanned flights, then to monkeys, then to one man, then to many, by incorporating *inside* the Space Centre more and more *outside* features brought back to the centre by each trial. The slow and progressive extension of a network from Cape Canaveral to the orbit of the earth is more of an achievement than the 'application' of calculations done inside the Space Centre to the outside world.

'Still, is not the application of science outside of the laboratories the best proof of its efficacy, of the quasi-supernatural power of scientists? Science *works* outside and its *predictions* are fulfilled.' Like all the other claims we have encountered in this chapter they are based on no independent and detailed study. No one has ever observed a fact, a theory or a machine that could survive *outside* of the networks that gave birth to them. Still more fragile than termites, facts and machines can travel along extended galleries, but they cannot survive one minute in this famous and mythical 'out-thereness' so vaunted by philosophers of science.

When the architects, urbanists and energeticians in charge of the Frango-castello solar village project in Crete had finished their calculations in early 1980 they had in their office, in Athens, a complete paper scale model of the village.[19] They knew everything available about Crete: solar energy, weather patterns, local demography, water resources, economic trends, concrete structures and agriculture in greenhouses. They had rehearsed and discussed every possible configuration with the best engineers in the world and had triggered the enthusiasm of many European, American and Greek development banks by settling on an optimal and original prototype. Like Cape Canaveral engineers they had *simply* to go 'out there' and apply their calculations, proving once again the quasi-supernatural power of scientists. When they sent their engineers from Athens to Frangocastello to start expropriating property and smoothing out the little details, they met with a totally unexpected 'outside'. Not only were the inhabitants not ready to abandon their lands in exchange for houses in the new village, but they were ready to fight with their rifles against what they took as a new American atomic military base camouflaged under a solar energy village. The application of the theory became harder every day as the mobilisation of opposition grew in strength, enrolling the pope and the Socialist Party. It soon became obvious that, since the army could not be sent to force Cretans to occupy willingly the future prototype, a negotiation had to start between the inside and the outside. But how could they strike a compromise between a brand new solar village and a few shepherds who simply wanted three kilometres of asphalted road and a gas station? The compromise was to abandon the solar village altogether. All the planning of the energeticians was routed back *inside* the network and limited to a paper scale model, another one of the many projects engineers have in their drawers. The 'out-thereness' had given a fatal blow to this example of science.

So how is it that in some cases science's predictions are fulfilled and in some other cases pitifully fail? The rule of method to apply here is rather straightforward: every time you hear about a successful application of a science, look for the progressive extension of a network. Every time you hear about a failure of science, look for what part of which network has been punctured. I bet you will always find it.

There was nothing more dramatic at the time than the prediction solemnly made a month in advance by Pasteur that on 2 June 1881 all the non-vaccinated sheep of a farm in the little village of Pouilly-le-Fort would have died of the terrible anthrax disease and that all the vaccinated ones would be in perfect health. Is this not a miracle, as if Pasteur had travelled in time, and in the vast world outside, anticipating a month in advance what will happen in a tiny farm in Beauce?[20] If, instead of gaping at this miracle, we look at how a network is extended, sure enough we find a fascinating negotiation between Pasteur and the farmers' representatives on how to *transform the farm into a laboratory*. Pasteur and his collaborators had already done this trial several times inside their lab, reversing the balance of forces between man and diseases, creating artificial epizootics in their lab (see Chapter 3). Still, they had never done it in full-scale farm conditions. But they are not fools, they know that in a dirty farm thronged by hundreds of onlookers they will be unable to repeat exactly the situation that had been so favourable to them (and will meet the same sort of failure as the energeticians bringing their village to the Cretans). On the other hand, if they ask people to come to *their* lab no one will be convinced (any more than telling Kennedy that Shepard has flown on the centrifuge one more time will convince the American people that they had taken their revenge over the Russians for being first in space). They have to strike a compromise with the organisers of a field test, to transform enough features of the farm into laboratory-like conditions – so that the same balance of forces can be maintained – but taking enough risk – so that the test is realistic enough to count as a trial done outside. In the end the prediction is fulfilled but it was in effect a *retro-diction*, exactly like the foresight of Professor Bijker on the future of Rotterdam harbour (see Part A) was in effect *hindsight*. To say this is not to diminish the courage of Shepard in his rocket, of the energeticians mobbed by the farmers, or of Pasteur taking the risk of a terrible mistake, any more than knowing in advance that Hamlet will die at the end of the play diminishes the talent of the actor. No amount of rehearsals frees the talented player from stage fright.

The predictable character of technoscience is entirely dependent on its ability to spread networks further. As soon as the outside is really encountered, complete chaos ensues. Of all the features of technoscience, I find this ability to extend networks and to travel along inside them the most interesting to follow; it is the most ingenious and the most overlooked of all (because of the inertia model depicted at the end of Chapter 3). Facts and machines are like trains, electricity, packages of computer bytes or frozen vegetables: they can go everywhere as long as the track along which they travel is not interrupted in the slightest. This dependence and fragility is not felt by the observer of science because 'universality' offers them the possibility of applying laws of physics, of biology, or of mathematics everywhere *in principle*. It is quite different *in practice*. You could say that it is possible in principle to land a Boeing 747 anywhere; but try in practice to land one on 5th Avenue in New York. You could say that telephone gives you a universal reach in principle. Try to call from San Diego someone in the middle of Kenya who does not, in practice, have a telephone. You can very well claim that Ohm's law (Resistance = Voltage/Current – see page 238) is universally applicable in principle; try in practice to demonstrate it without a voltmeter, a wattmeter and an ammeter. You may very well claim that in principle a navy helicopter can fly anywhere; but try to fix it in the Iranian desert when it is stalled by a sandstorm, hundreds of miles from the aircraft carrier. In all these mental experiments you will feel the vast difference between principle and practice, and that when everything works according to plan it means that you do not move an inch out of well-kept and carefully sealed networks.

Every time a fact is verified and a machine runs, it means that the lab or shop conditions have been extended *in some way*. A medical doctor's cabinet a century ago would have been furnished with an armchair, a desk and maybe an examination table. Today, your doctor's cabinet is filled with dozens of instruments and diagnostic kits. Each of them (like the thermometer, the blood pressure kit or the pregnancy test) has come from a laboratory to the cabinet through the instrument industry. If your doctor verifies the application of the laws of physiology, very well, but do not ask her to verify them in an empty cabin in the middle of the jungle, or she will say, 'Give me my instruments back first!' Forgetting the extension of the instruments when admiring the smooth running of facts and machines would be like admiring the road system, with all those fast trucks and cars, and overlooking civil engineering, the garages, the mechanics and the spare parts. Facts and machines have no inertia of their own (Chapter 3); like kings or armies they cannot travel without their retinues or impedimenta.

(2) Tied in by a few metrological chains

The dependency of facts and machines on networks to travel back from the centres to the periphery makes our job much easier. It would have been impossible for us to follow 'universal' laws of science that would have been applicable everywhere without warning. But the progressive extension of the domain of application of a laboratory is very simple to study: just follow the *traces* this application creates. As we saw in Part B, a calculation on paper can apply to the outside world only if this outside world is itself another piece of paper of the same format. At first, this requirement seems to mark the end of the road for the calculations. It is impossible to transform Sakhalin, Rotterdam, turbulences, people, microbes, electrical grids and all the phenomena out there into a paper world similar to the one in there. This would be without allowing for the ingenuity of the scientists in extending everywhere the instruments that produce this paper world. **Metrology** is the name of this gigantic enterprise to make of the outside a world inside which facts and machines can survive. Termites build their obscure galleries with a mixture of mud and their own droppings; scientists build their enlightened networks by giving the outside the same paper form as that of their instruments inside. In both cases the result is the same: they can travel very far without ever leaving home.

In the pure, abstract and universal world of science the extension of the new objects created in the labs costs nothing at all. In the real, concrete and local world of technoscience, however, it is frightfully expensive simply to maintain the simplest physical parameters stable. A simple example will be enough. If I ask, 'What time is it?' you will have to look at your watch. There is no way to settle this question without taking a *reading* at the window of this scientific instrument (the sun will do, but not when you need to catch a train). No matter how humble it is, the clock is of all scientific instruments the one with the longest and most influential history. Remember that Lapérouse carried with him no less than twelve ship chronometers and had several scientists on board simply to check and compare their movements. His whole trip would have been rendered useless if he could not have kept the time constant. Now, if our two watches disagree, we will be led to a third one which will act as our referee (a radio station, a church clock). If there is still a disagreement on the quality of the clock used as referee, we might very well call the 'speaking clock'. If one of us was as obstinate as the dissenter of Chapters 1 and 2, he or she will be led into an extraordinarily complex maze of atomic clocks, lasers, satellite communications: the International Bureau of Time coordinating throughout the earth what time it is. Time is not universal; every day it is made slightly more so by the extension of an international network that ties together, through visible and tangible linkages, each of all the reference clocks of the world and then organises secondary and tertiary chains of references all the way to this rather imprecise watch I have on my wrist. There is a continuous trail of readings, checklists, paper forms, telephone lines, that tie all the clocks together. As soon as you leave this trail, you start to be *uncertain* about what time it is, and the only way to regain certainty is to get in touch again with the metrological chains. Physicists use the nice word *constant* to designate these elementary parameters necessary for the simplest equation to be written in the laboratories. These constants, however, are so inconstant that the US, according to the National Bureau of Standards, spends 6 per cent of its Gross National Product, that is, three times what is spent on R & D (see Chapter 4), just to maintain them stable![21]

That much more effort has to be invested in extending science than in doing it may surprise those who think it is naturally universal. In the figures that I presented in Chapter 4 we could not make sense at first of this mass of scientists and engineers engaged in management of R & D, management, inspection, production, and so on (see page 16). It need no longer surprise us. We know that scientists are too few to account for the enormous effect they are supposed to generate and that their achievements circulate in frail, recent, costly and rare galleries. We know that 'science and technology' is only the abstracted tip of a much larger process, and has only a very vague resemblance to it. The paramount importance of metrology (like that of development and industrial research) gives us a measure, so to speak, of our ignorance.

These long metrological chains necessary for the very existence of the simplest laboratories concern only the official constants (time, weight, length, biological standards, etc.), but this is only a tiny part of all the measurements made. We are so used to the pervasive presence of all these meters, counters, paper forms and tallies which pave the way for centres of calculation that we forget to consider each of them as the sure trace of an earlier *invasion* by a scientific profession. Just think about the kind of answer you can provide to these questions: How much did I earn this month? Is my blood pressure above or below normal? Where was my grandfather born? Where is the tip of Sakhalin island? How many square metres is my flat? How much weight have you put on? How many good grades did my daughter get? What temperature is it today? Is this pack of beer on sale a good buy? Depending on who asks these questions you may provide either a *softer* answer or a *harder* one. In the latter case you will have to fall back on a paper form: the accounting slip sent to you by your bank; the reading taken out of the blood pressure kit in your doctor's office; the birth certificates kept at City Hall or a genealogical tree; the list of flashing lights printed in the *Nautical Almanac*; a geometrical drawing of your flat; a scale; a school report kept in your daughter's college administration; a thermometer; the dozens of metrological marks made on the pack of beer (content, alcoholic degree, amount of preservatives, etc.). What we call 'thinking with accuracy' in a situation of controversy is always bringing to the surface one of these forms. Without them we simply *don't know*.

If for one reason or another (crime, accident, controversy), the dispute is not settled at this point, you will be led along one of the many metrological chains that pile up paper forms to the *nth* order. Even the question 'who are you' cannot be solved, in some extreme situations, without superimposing passports to fingerprints to birth certificates to photographs, that is without constituting a *file* that brings together many different paper forms of various origins. *You* might very well know who you are and be satisfied with a very soft answer to this absurd query, but the policeman, who raises the question from the point of view of a centre, wants to have a harder answer than that, exactly as when Lapérouse kept asking the Chinese fishermen where they were in terms of longitude and latitude. We can understand now the misunderstanding studied in Chapter 5, Part C between the softer and the harder ways of solving the paradox of the fact-builder. The requirements put on knowledge are utterly different if one wants to use it to settle a local dispute or to participate in the *extension* of a network far away. All the intermediaries are enough in the first case (I know who I am, what time it is, if it is warm or cold, if my flat is big or small, if I earn enough, if my daughter works well, if Sakhalin is an island or not). They are all found *wanting* in the latter case. The misunderstanding is of the same nature and has the same concrete meaning as if an army engineer in charge of preparing the landing of *B52* bombers on a Pacific island finds only a muddy landing strip a few hundred yards long. He will indeed be disappointed and will find the airstrip wanting.

The only way to prepare 'landing strips' everywhere for facts and machines is to transform as many points as possible of the outside world into instruments. The walls of the scientific galleries are literally *papered over*.

Machines, for instance, are drawn, written, argued and calculated, before being built. Going from 'science' to 'technology' is not going from a paper world to a messy, greasy, concrete world. It is going from paperwork to still more paperwork, from one centre of calculation to another which gathers and handles more calculations of still more heterogeneous origins.[22] The more modern and complex they are, the more paper forms machines need so as to come into existence. There is a simple reason for this: in the very process of their construction they disappear from sight because each part hides the other as they become darker and darker black boxes (Chapter 3). The *Eagle* group, during the debugging, had to build a computer program just to keep track of the modifications each of them was doing to the prototype, just to remember what *Eagle* was about, to keep it synoptically under their eyes while it became more and more obscure (Introduction). Of all the parts of technoscience, the engineers' drawings and the organisation and management of the traces generated simultaneously by engineers, draughtsmen, physicists, economists, accountants, marketing agents and managers are the most revealing. They are the ones where the distinctions between science, technology, economics and society are the most absurd. The centres of calculations of major machine-building industries concentrate on the same desks paper forms of all origins, recombining them in such a way that some slips of paper bring together the shape of the part to be built (drawn in a codified geometrical space); the tolerance and calibration necessary for its construction (all the metrological chains inside and outside the forms); the physical equations of material resistance; the names of the workers in charge of the parts; the mean time necessary to effect the operations (result of decades of taylorisation); the dozens of codes that make the keeping of the inventory possible; the economic calculations; and so on. Those who would try to replace the common history of these centres of calculations by clean distinct histories of science, of technology, and of management would have to butcher the subject.

Each of these paper forms is necessary for one of the dozens of sciences involved in machine-building simply to have any relevance at all. Accountancy, for instance, is a crucial and pervasive science in our societies. Its extension, however, is strictly limited by the few paper forms that make accurate book-keeping possible. How do you apply book-keeping to the confusing world of goods, consumers, industry? Answer: by transforming each of these complex activities, so that, at one point or another, they generate a paper form that is readily applicable to book-keeping. Once each hamburger sold in the United States, each coffee cup, each bus ticket is accompanied by a numbered stub, or one of these little white tallies spews out of every cash register, then indeed accountants, managers and economists are able to expand their skill at calculating. A restaurant, a supermarket, a shop, an assembly line are generating as many readings from as many instruments as a laboratory (think of the scales, the clocks, the registers, the order forms). It is only once the *economy* is made to generate enough of these paper forms so as to resemble *economics* that the economists become part of an expanding profession. There is no reason to limit the study of science to the writing of the Book of Nature, and to forget to study this 'Great Book of Culture' which has a much more pervasive influence on our daily life than the other – the mere information in banks, for example, is several orders of magnitude more important than scientific communication.

Even geography, that seems so readily applicable 'outside', once the map is made, cannot escape very far from the networks without becoming useless. When we use a map, we rarely compare what is written on the map with the landscape – to be capable of such a feat you would need to be yourselves a well-trained topographer, that is, to be *closer* to the profession of geographer. No, we most often *compare* the readings on the map with the road *signs* written in the *same* language. The outside world is fit for an application of the map only when all its relevant features have themselves been written and marked by beacons, landmarks, boards, arrows, street names and so on. The easiest proof of this is to try to navigate with a very good map along an unmarked coast, or in a country where all the road boards have been torn off (as happened to the Russians invading Czechoslovakia in 1968). The chance is that you will soon be wrecked and lost. When the out-thereness is really encountered, when things out there are seen for the *first* time, this is the end of science, since the essential cause of scientific superiority has vanished.

The history of technoscience is in a large part the history of all the little inventions made along the networks to accelerate the mobility of traces, or to enhance their faithfulness, combination and cohesion, so as to make action at a distance possible. This will be our **sixth principle**.

(3) About a few other paper-shufflers

If we extend the meaning of metrology to include not only the upkeep of the basic physical constants but also the transformation of as many features as possible of the outside in paper forms, we might end up studying the most despised of all the aspects of technoscience: the paper-shufflers, the red-tape worms, the bureaucrats. Ah! these bureaucrats, how hated they are – these people who only deal with pieces of paper, files and forms, who know nothing about the real world, but are only superimposing forms on other forms simply to check if they have been correctly filled in; this curious breed of lunatics that prefers to believe a piece of paper to any other source of information, even if it is against common sense, logic and even their own feelings. Sharing this scorn would be, however, a major mistake for us who wish to follow science in action up to the end. First, because what are seen as defects in the case of the paper-shufflers are considered noble qualities when considering these other paper-shufflers who are called scientists and engineers. Believing more the *nth* order paper form than common sense is a feature of astronomers, economists, bankers, of everyone who treats in the centres phenomena which are, by definition, absent.

It would be a mistake, second, because it is through bureaucracy and inside the files that the results of science travel the furthest. For instance, the loggings produced by *Schlumberger* engineers on oil platforms (Part A, section 2) become part of a file inside a bank at Wall Street that combines geology, economics, strategy and law. All these unrelated domains are woven together once they become sheets of this most despised of all objects, the *record*, the dusty record. Without it, though, the loggings would stay where they were, inside the *Schlumberger* cabin or truck, without any relevance to other issues. The microbiological tests of water made by bacteriologists would have no relevance either if they stayed inside the lab. Now that they are integrated, for instance, in another complex record at City Hall that juxtaposes architects' drawings, city regulations, poll results, vote tallies and budget proposals, they profit from each of these other skills and crafts. Understanding the bearing of bacteriology on 'society' might be a difficult task; but following in how many legal, administrative and financial operations bacteriology has been enrolled is feasible: just follow the trail. As we saw in Chapter 4, the esoteric character of a science is inversely proportional to its exoteric character. What we realise now is that administration, bureaucracy, and management in general are the only big resources available to expand really far: the government supports the bacteriology laboratory which has become an obligatory passage point for every decision to be made. What appeared at the beginning of this book as vast and insulated pockets of science are probably best understood if they are seen to be scattered through centres of calculation, dispersed over files and records, seeded through all the networks and visible only because they accelerate the local mobilisation of some resources among many others that are necessary to administer many people on a large scale and at a distance.[23]

The third and final reason why we should not despise bureaucrats, managers, paper-shufflers or, in brief, this tertiary sector that completely dwarfs the size of technoscience is that it constitutes a mixture of other disciplines which have to be studied with the *same* method I have presented in this book even though they are not considered as pertaining to 'science and technology'. When people claim they want to explain 'socially' the development of 'science and technology' they use entities like national policy, multinational firms' strategies, classes, world economic trends, national cultures, professional status, stratification, political decisions, and so on and so forth. At no point in this book have I used any of these entities; on the contrary, I have explained several times that we should be as agnostic about society as about nature, and that providing a social explanation does not mean anything 'social' but only something about the relative solidity of as*socia*tions. I also promised, however, at the end of Chapter 3, that we will meet at some point a stable state of society. Well, here we are: a stable state of society is produced by the multifarious administrative sciences exactly like a stable interpretation of black holes is provided by astronomy, of microbes by bacteriology, or of proven oil reserves by geology. No more, no less. Let us end with a few more examples.

The state of the economy, for instance, cannot be used unproblematically to explain science, because it is itself a very controversial outcome of another soft science: economics. As we saw earlier, it is extracted out of hundreds of statistical institutions, questionnaires, polls and surveys, and treated in centres of calculation. Something like the Gross National Product is an *nth* order visual display which, to be sure, may be combined to other paper forms, but which is no more *outside* the frail and tiny networks built by economists than stars, electrons or plate tectonics. The same is true for many aspects of politics. How do we know that Party A is stronger than Party B? Each of us may have an opinion about the relative strength of these parties; indeed, it is because each of us has one opinion about it that we may have to build a huge scientific experiment to settle the question. Scientific? Sure. What is a national election, if not the transformation through a very costly and cumbersome instrument of all the opinions into marks on ballot papers, marks which are then counted, summed, compared (with great care and with much controversy) to eventually end up in one *nth* order visual display: Party A: 51%, Party B: 45%, Null: 4%? To distinguish between or oppose science, politics and economics would be meaningless from our point of view, because in terms of size, relevance and cost, the few figures that decide the Gross National Product or the political balance of forces are much more important, trigger much more interest, much more scrutiny, much more passion, much more scientific method than a new particle or a new radio source. All of them depend on the same basic mechanism: calibrating inscription devices, focusing the controversies on the final visual display, obtaining the resources necessary for the upkeep of the instruments, building *nth* order theories on the archived records. No, our method would gain nothing in explaining 'natural' sciences by invoking 'social' sciences. There is not the slightest difference between the two, and they are both to be studied in the same way. Neither of them should be believed more nor endowed with the mysterious power of jumping out of the networks it builds.

What is clear for economics, politics and management is all the clearer for sociology itself. How could someone who decided to follow scientists in action forget to study sociologists striving to define what society is all about, what keeps us all glued together, how many classes there are, what is the aim of living in society, what are the major trends of its evolution? How could one believe these people who say what society is about more than the others? How could one transform astronomers into spokespersons for the sky and still accept that the sociologists tell us what society *is*. The very definition of a 'society' is the final outcome, in Sociology Departments, in Statistical Institutions, in journals, of other scientists busy at work gathering surveys, questionnaires, archives, records of all sorts, arguing together, publishing papers, organising other meetings. Any agreed definition marks the happy end of controversies like all the settlements we have studied in this book. No more, no less. The results on what society is made of do not spread more or faster than those of economics, topology or particle physics. These results too would die if they went outside of the tiny networks so necessary for their survival. A sociologist's interpretation of society will not be substituted for what every one of us thinks of society without *additional* struggle, without textbooks, chairs in universities, positions in the government, integration in the military, and so on, exactly as for geology, meteorology or statistics.

No, we should not overlook the administrative networks that produce, *inside* rooms in Wall Street, in the Pentagon, in university departments, fleeting or stable representations of what is the state of the forces, the nature of our society, the military balance, the health of the economy, the time for a Russian ballistic missile to hit the Nevada desert. To rely on social sciences more than on natural ones would put our whole journey in jeopardy, because we would have to accept that the space-time elaborated *inside* a network by *one* science has spread outside and *included* all the others. We are no more included in the space of society (built by sociologists through so many disputes), than in the time of geology (slowly elaborated in Natural History Museums), or in the domain of neurosciences (carefully extended by neuroscientists). More exactly, this inclusion is not naturally provided without additional work; it is obtained locally if the networks of sociologists, geologists and neuroscientists are extended, if we have to pass through their laboratories, or through their metrological chains, if they have been able to render themselves indispensable to our own trips and travels. The situation is exactly the same for the sciences as for gas, electricity, cable TV, water supplies or telephones. In all cases you need to be hooked up to costly networks that have to be maintained and extended. This book has been written to provide a breathing space to those who want to study independently the extensions of all these networks. To do such a study it is absolutely necessary never to grant to any fact, to any machine, the magical ability of leaving the narrow networks in which they are produced and along which they circulate. This tiny breathing space would become immediately vitiated if the same fair and symmetric treatment was not applied to the social and administrative sciences as well.

APPENDIX 1

Rules of Method

Rule 1 We study science *in action* and not ready made science or technology; to do so, we either arrive before the facts and machines are blackboxed or we follow the controversies that reopen them. (Introduction)

Rule 2 To determine the objectivity or subjectivity of a claim, the efficiency or perfection of a mechanism, we do not look for their *intrinsic* qualities but at all the transformations they undergo *later* in the hands of others. (Chapter 1)

Rule 3 Since the settlement of a controversy is the *cause* of Nature's representation, not its consequence, we can never use this consequence, Nature, to explain how and why a controversy has been settled. (Chapter 2)

Rule 4 Since the settlement of a controversy is the *cause* of Society's stability, we cannot use Society to explain how and why a controversy has been settled. We should consider symmetrically the efforts to enrol human and non-human resources. (Chapter 3)

Rule 5 We have to be as *undecided* as the various actors we follow as to what technoscience is made of; every time an inside/outside divide is built, we should study the two sides simultaneously and make the list, no matter how long and heterogeneous, of those who do the work. (Chapter 4)

Rule 6 Confronted with the accusation of irrationality, we look neither at what rule of logic has been broken, nor at what structure of society could explain the distortion, but to the angle and direction of the observer's *displacement*, and to the *length* of the network thus being built. (Chapter 5)

Rule 7 Before attributing any special quality to the mind or to the method of people, let us examine first the many ways through which inscriptions are gathered, combined, tied together and sent back. Only if there is something unexplained once the networks have been studied shall we start to speak of cognitive factors. (Chapter 6)

258

APPENDIX 2

Principles

First principle The fate of facts and machines is in later users' hands; their qualities are thus a consequence, not a cause, of a collective action. (Chapter 1)

Second principle Scientists and engineers speak in the name of new allies that they have shaped and enrolled; representatives among other representatives, they add these unexpected resources to tip the balance of force in their favour. (Chapter 2)

Third principle We are never confronted with science, technology and society, but with a gamut of weaker and stronger *associations*; thus understanding *what* facts and machines are is the same task as understanding *who* the people are. (Chapter 3)

Fourth principle The more science and technology have an esoteric content the further they extend outside; thus, 'science and technology' is only a subset of technoscience. (Chapter 4)

Fifth principle Irrationality is always an accusation made by someone building a network over someone else who stands in the way; thus, there is no Great Divide between minds, but only shorter and longer networks; harder facts are not the rule but the exception, since they are needed only in a very few cases to displace others on a large scale out of their usual ways. (Chapter 5)

Sixth principle History of technoscience is in a large part the history of the resources scattered along networks to accelerate the mobility, faithfulness, combination and cohesion of traces that make action at a distance possible. (Chapter 6)

Notes

Introduction

1 I am following here James Watson's account (1968).
2 I am following here Tracy Kidder's book (1981). This book, like Watson's, is compulsory reading for all of those interested in science in the making.
3 On this episode see T.D. Stokes (1982).
4 This notion of under-determination is also called the Duhem-Quine principle. It asserts that no one single factor is enough to explain the closure of a controversy or the certainty acquired by scientists. This principle forms the philosophical basis of most social history of sociology of science.

Chapter 1

1 This debate about the MX weapon system has been the object of a long public controversy in the USA.
2 This example is taken from Nicholas Wade (1981). The rest of the controversy is inspired from the book, although it is in part fictional.
3 This example is taken from Michel Callon (1981).
4 Cited in S. Drake (1970, p. 71).
5 I am using here the following article: A. V. Schally, V. Baba, R. M. G. Nair, C. D. Bennett (1971), 'The amino-acid sequence of a peptide with growth hormone-releasing isolated from porcine hypothalamus', *Journal of Biological Chemistry*, vol. 216, no. 21, pp. 6647–50.
6 The field of citation studies has become an independent sub-discipline. For a review see E. Garfield (1979) or the review *Scientometrics* for more recent and more specialised examples. For the context of citation, see M. H. MacRoberts and B. R. MacRoberts (1986).
7 This expression has become traditional since the work of Thomas Kuhn (1962).
8 The Science Citation Index is produced by the Institute for Scientific Information in Philadelphia and has become the basis of much work in science policy.
9 I am using here the following article: R. Guillemin, P. Brazeau, P. Böhlen, F. Esch, N. Ling, W. B. Wehrenberg (1982), 'Growth-hormone releasing

factor from a human pancreatic tumor that caused acromegaly', *Science*, vol. 218, pp. 585-7.

10 The article commented on here is by C. Packer, 'Reciprocal altruism in papio P.', *Nature* 1977 Vol. 265, no. 5593, pp. 441-443. Although this transformation of the literature is a sure telltale of the differences between harder and softer fields, I know of no systematic study of this aspect. For a different approach and on the articles in physics see C. Bazerman (1984).

11 See M. Spector, S. O'Neal, E. Racker (1980), 'Regulation of phosporylation of the ß-subunit of the Ehrlich Ascites tumor $Na^\rightarrow K^\rightarrow$-ATPase by a protein kinase cascade'. *Journal of Biological Chemistry*, vol. 256, no. 9 pp. 4219-27. On this and many other borderline cases, see W. Broad and N. Wade (1982).

12 For a general presentation see M. Callon, J. Law and A. Rip (eds) (1986).

13 On the somatostatin episode see Wade (1981, chapter 13).

14 For a good introduction or rhetoric in settings other than the scientific ones see C. Perelman (1982).

Chapter 2

1 For an introduction to bibliometry and to the study of citations see E. Garfield (1979); for the co-words analysis see M. Callon, J. Law and A. Rip (eds) (1986); for an introduction to semiotics see F. Bastide (1985).

2 I am following here the work of Trevor Pinch (1986).

3 I am following here the work of Mary Jo Nye (1980, 1986).

4 On this see N. Wade (1981, Chapter 13).

5 I am following here the empirical example studied by H. Collins (1985), although his description of the ways of settling controversies is rather different and will be analysed in Part II of this book.

6 I am following here the work of Farley and Geison, (1974).

7 Later on, however, the controversy was resumed; see R. Dubos (1951). There are always only practical and temporary ends to controversies, as will be shown in the last section.

8 On this controversy see M. Mead (1928) and D. Freeman (1983).

9 I am using here D. MacKenzie's (1978) article. See also his (1981) book on the larger setting of the same controversy.

10 On this episode of the discovery of somatostatin see N. Wade (1981 chapter 13).

11 This excerpt is taken from E. Duclaux's *Traité de biochimie* (1896), vol. II, p. 8. Duclaux was collaborator of Pasteur.

12 I am using here the following article by Pierre and Marie Curie: (1898) 'Sur une substance nouvelle radio-active, contenue dans la pechblende', *Comptes Rendus de l'Académie des Sciences*, vol. 127, pp. 175-8.

13 For the definition of these words and of all the concepts of semiotics see A. Greimas and J. Courtès (1979/1983). For a presentation of semiotics in English see F. Bastide (1985).

14 See J. W. Dauben (1974).

15 For the ultracentrifuge see the nice study by Boelic Elzen (forthcoming).

16 I am alluding here to the remarkable work by A. Desmond (1975).

17 This basic question of relativism has been nicely summed up in many articles by Harry Collins. See in particular his latest book (1985).

Chapter 3

1 For a presentation of laboratory studies see K. Knorr (1981), K. Knorr and M. Mulkay (eds) (1983) and M. Lynch (1985).

2 I am following in this introduction the article by L. Bryant (1976); see also his (1969) article.

3 On this controversy see again D. Freeman (1983) and on the general history surrounding this episode see D. Kevles (1985).

4 I am following here J. Geison's article on Pasteur (1974).

5 On this dramatic episode see R. Dubos & J. Dubos (1953).

6 I follow here T.P. Hughes (1971).

7 On this see D. Kevles (1978), on the many different strategies to interest a society in the development of a profession.

8 This knowledge seems excessive to many sociologists of science (see S. Woolgar (1981), M. Callon and J. Law (1982), B. Hindess (1986)), and seems quite reasonable to the founder of the interest theory Barry Barnes (1971), to D. Bloor (1976), and to S. Shapin (1982).

9 See L. Szilard (1978, p. 85).

10 I am using here R. Jenkins's article (1975).

11 See B. Rozenkranz (1972) and D. Watkins (1984)

12 See M. Callon (1981).

13 On this notion of 'idea' see the last part of this chapter.

14 This example is taken from L. Tolstoy's masterpiece (1869).

15 This expression has been proposed by J. Law (1986) in correlation with his notion of 'heterogeneous engineering'.

16 On this, see the notion of 'reverse salient' proposed by T. Hughes (1983).

17 I am using here L. Hoddeson's article (1981).

18 I follow here S. Shapin (1979).

19 On this and the following see A. Leroi-Gourhan (1964).

20 The traditional difference between human – those who are able to speak and are endowed with wills – and non-human – those supposed to be mute and denied wills and desires – is immaterial here and is not enough to break the necessary symmetry. On this see M. Callon (1986).

21 On Newcomen's engine see B. Gille (1978)

22 For a reader, a bibliography and an introduction to these many strategies, see D. MacKenzie and J. Wajcman (1985).

23 For a critical introduction to the notion of discovery, see A. Brannigan (1981).

24 Defined by David Bloor in his classic book (1976) and to which he opposes his principle of symmetry that requires an explanation to apply in the same terms to winners and losers.

25 This example and many others are sketched in the non-technical book written by T. Peters and N. Austin (1985).

Chapter 4

1 I follow here Roy Porter's account (1982). See also his (1977) book on the formation of the new discipline of geology.

2 See D. Kevles (1978) as an excellent example of the historical study of a scientific profession.

3 This example is a collage.

4 Although all the elements are accurate, this is an ideal-type and not a real example.

5 See T. Kidder (1981).

6 Most of the figures used in this part come from the National Science Foundation's *Science Indicators* published in Washington every two years.

7 See OECD (1984).

8 Number of doctorates in the US: total: 360,000; in research: 100,000; in development: 18,000 (*Science Indicators*, 1983, p. 254).

9 Number of scientists and engineers engaged in R & D by type of occupation and employer in the US:

Engaged in research:
355,000; of these 98,000 are in industry; the rest in universities or in Federal labs;
Engaged in development:
515,000; of these 443,000 are in industry; the rest in Universities or in Federal Laboratories;
Engaged in management of R & D: 224,000; of these 144,000 are in industry; the rest 224,000; of these 144,000 are in industry; the rest in universities or in Federal labs.

Science Indicators 1982 1983, p. 277)

10 Number of US doctorate scientists doing R & D apart from those in business and industry:

Basic science:	48,000
Applied research:	24,500
Development	2,900
Management of R & D	13,800

(*SI*, 1983, p. 311)

11 On this long-term, large-scale trend see D. de S. Price (1975); see also N. Rescher (1978).

12 On the notion of stratification see the classic study by J. and S. Cole (1973).

13 On visibility and on the many other notions developed by the American School of sociology of scientists and engineers – in contradistinction to the sociology of science and technology mostly used in this book – see the classic book of K. Merton (1973).

14 Comparative shares of research institutions in R & D budget in the US:

Top 10 doing 20%   Top 100 doing 85%

Science Indicators 1982, 1983, p. 125)

15 Comparative shares of the six top Western countries in the R & D budget, literature, patents, and citations:

US proportion of the world's science and technology articles: 37%
(in the lowest field, chemistry, it is 21%; in the highest, biomedicine, it is 43%)

(SI, 1982, p. 11)

US proportion of the Western world's budget spent on R & D: 48% in 1979
(Japan 15%; European Community 30%)

(OECD 1983, p. 21)

US proportion of the Western world's workforce in R & D: 43% in 1979
(Japan 26%; European Community 27%)

(idem)

16 This situation of dependence is much worse if we consider not only the top industrial countries but also the smaller ones or the underdeveloped countries. When we take the poorest countries into account, what is officially defined as technoscience vanishes from view. Determining its scale is no longer the right expression. We should now talk in terms of *traces*. A few institutes staffed for the most part with scientists from the developed countries are almost invisible, scattered among the hundreds of millions who know nothing about the interior of facts and machines. See the figures in UNESCO (1983).

17 See on this notion of mobilisation W. McNeill's major book (1982) and Chapter 6.

Chapter 5

1 See David Bloor (1976). On this debate see M. Hollis and S. Lukes (1982) and E. Mendelsohn and Y. Elkana (1981). The two most interesting articles on this debate are without doubt those of R. Horton (1967; 1982).

2 This example from E. E. Evans-Pritchard's classic book (1937) has been turned into a canonic topic for anthropology of science by David Bloor (1976).

3 This example is taken from Edward Hutchins (1980).

4 I am following here D.A. Hounshell (1975).

5 See on this succession of contradictory accusations B. Easlea (1980).

6 See on this point B.J.T. Dobbs (1976).

7 This is an adaptation of D. Bloor's drawing (1976, p. 126).

8 Naturally, I am following here the canonic example offered by Bloor and not the very subtle interpretations offered by Evans-Pritchard.

9 See on this point the classic book edited by B. Wilson (1970).

10 I am following here M. Cole and S. Scribner (1974); other examples by A.R. Luria are to be found in his (1976) book edited by M. Cole.

11 This other canonic example is taken from R. Bulmer (1967) and has been treated at length by B. Barnes (1983).

12 The most complete work of ethnoscience is to be found in H. Conklin (1980). Unfortunately there is no equivalent of this on a Western industrialised community.

13 I am using here the beautiful book of A. Desmond (1975), especially the chapter 6.

14 This example is taken from M. Callon (1986).

15 His testimonies form the bulk of M. Augé's book (1975). For obvious reasons, Augé never published the result of the corpse interrogation of his friend.

16 This example is taken from J. Gusfield's book which is a unique case of anthropology of belief/knowledge in a modern Western society (1981).

17 This is why 'oral cultures' have been thought to be both rigid and devoid of innovation. On this see J. Goody's pioneering work (1977).

18 On this transformation and transportation of other people's beliefs see P. Bourdieu (1972/1977) J. Fabian (1983) and the recent book on field trips edited by G.W. Stocking (1983).

Chapter 6

1 On this episode see J.-F. Lapérouse (no date) and F. Bellec (1985).

2 I am following here J. Law's account of this episode (1986). On all this redefinition of capitalism in terms of long distance networks the essential work is of course that of F. Braudel (1979/1985).

3 The literature on expeditions and collections is not very extensive but there are some interesting case studies. Among them are L. Brockway (1979) and L. Pyenson (1985).

4 This example is taken from L. Allaud et M. Martin (1976).

5 I follow here E. Eisenstein's account (1979). Her book is essential reading for all of those who wish, as she says, to 'reset the stage for the Copernican Revolution'.

6 For a general review of this question see the volume I edited in French with J. de Noblet (1985).

7 On this comparison between botanists and ethnobotanists see H. Conklin (1980).

8 I follow here B. Bensaude-Vincent's account (1986). See also her thesis (1981) and on Mendeleev's work see F. Dagognet (1969).

9 Actually, the strength of the table came later from the unexpected correspondence between the classification and the atomic theory that retrospectively explained it.

10 This example is elaborated in M. Polanyi (p. 83).

11 For an interesting study see that of F. Fourquet (1980) on the construction of INSEE, the French institution gathering statistics.

12 See P.S. Stevens (1978). On this question of the relations between scale models, models and calculations, probably the best book is still M. Black (1961). Less known but very useful is the work of F. Dagognet. See in particular is recent book (1984).

13 I am following here the exemplary article of T. Hughes (1979).

14 This useful word has been proposed by E. Gerson and L. Star to describe much the same mechanism as the one I name here 'cascade'. This chapter owes much to the work of their Tremont Institute in California.

15 This does not mean that 'theories' simply follow the accumulation of 'data' – on the contrary 'mere stamp collecting' is often opposed to 'real science' – but simply that any a priori epistemological distinction between the two makes the study impossible. The problem is that we lack independent studies on the construction of this contrast between 'data' and 'theories'. For such an endeavour on the relations between physics and chemistry see I. Stengers (1983).

16 See on this A. Koyré (1966) and S. Drake (1970).

17 This has to be taken with a grain of salt since there is no study pertaining to anthropology of science which tackles this question. A related effort is to be found in E. Livingston's recent book (1985).

18 I am using here the excellent book of T. Wolfe (1979). To the humiliation of our profession, we have to confess that some of the best books on technoscience, those of Kidder, Watson and Wolfe, for example, have not been written by professional scholars.

19 This example is taken from one of the rare long-term, empirical studies of a modern large-scale technical project by M. Coutouzis (1984); see also our article (1986) (Coutouzis and Latour).

20 On this episode see J. Geison (1974).

21 See the article by P. Hunter (1980).

22 Within the small but fascinating literature on this topic, the best introduction is the work of P.J. Booker (1979) and Baynes K. & Pugh F. (1981). For a shorter introduction see E. Ferguson (1977).

23 On this dispersion of the sciences as on so many microtechnics of power see M. Foucault's work, especially (1975).

References

Allaud, L. and M. Marttin (1976). *Schlumberger, Histoire d'une Technique.* Paris,
Augé, Marc (1975). *Théorie des pouvoirs et idéologie.* Paris, Hermann.
Barnes, Barry (1974). *Scientific Knowledge and Sociological Theory.* London, Routledge
& Kegan Paul.
  (1982). *T.S. Kuhn and Social Science.* London, Macmillan.
  (1983). 'On the conventional character of knowledge and cognition'. In K. Knorr and
  M. Mulkay (eds), pp.19–53.
Bastide, Françoise (1985). The semiotic analysis of scientific discourse. Paris, Ecole de
  mines, miméo.
Baynes, Ken and Pugh, Francis (1981). *The Art of the Engineer.* Guildford, Lutherwood
  Press.
Bazerman, Charles (1984). 'Modern evolution of the experimental report of physics:
  spectroscopic articles in *Physical Review'. Social Studies of Science*, vol. 14, no. 2,
  pp.163–97.
Bellec, Francois (1985). *La Généreuse et tragique expédition de Lapérouse.* Rennes, Ouest
  France.
Bensaude-Vincent, Bernadette (1981). *Les Pièges de l'élémentaire. Contribution à l'histoire
  de l'élément chimique.* Thèse de Doctorat. Université de Paris I.
  (1986). 'Mendeleev's periodic system of chemical elements' *British Journal for the
  History of Science*, vol. 19, pp. 3–17.
Black, Max (1961). *Models and metaphors*, Ithaca, Cornell University Press.
Bloor, David (1976). *Knowledge and Social Imagery.* London, Routledge & Kegan Paul.
Booker, P.J. (1979). *A History of Engineering Drawing.* London, Northgate.
Bourdieu, Pierre (1972/1977). *Outline of a Theory of Practice.* Cambridge. Cambridge
  University Press.
Brannigan, Augustine (1981). *The Social Basis of Scientific Discoveries.* Cambridge
  University Press.
Braudel, Fernand (1979/1985). *The Perspective of the World. 15th to 18th Century.* New
  York, Harper & Row.
Broad, William and Wade, Nicholas (1982). *Betrayers of the Truth: Fraud and Deceit in the
  Halls of Science*, New York, Simon & Schuster.
Brockway, Lucile H. (1979). *Science and Colonial Expansion: The Role of the British
  Royal Botanic Gardens.* New York, Academic Press.
Brown, Lloyd A. (1949/1977). *The Story of Maps.* New York, Dover.

266

Bryant, Lynwood (1969). 'Rudolf Diesel and his rational engine'. *Scientific American*, vol. 221, pp.108–17.

(1976). 'The development of the Diesel Engine'. *Technology and Culture*, vol. 17, no. 3, pp.432–46.

Bulmer, Ralph (1967). 'Why is a cassowary not a bird? A problem of zoological taxonomy among the Karam'.

Callon, Michel (1981). 'Struggles and negotiations to decide what is problematic and what is not: the sociologic'. In K. Knorr, R.K. Krohn & R. Whitley (eds). pp. 197–220.

Callon, Michel and Law, John (1982). 'On interests and their transformation: enrolment and counter-enrolment'. *Social Studies of Science*, vol. 12, no. 4, pp.615–26.

Callon Michel (1986). 'Some elements of a sociology of translation: domestication of the scallops and the fishermen'. John Law (editor), pp. 196–229.

Callon, Michel, Law, John, and Rip, Arie (eds) (1986). *Mapping the Dynamic of Science and Technology*. London, Macmillan.

Cole, J. and Cole, S. (1973). *Social Stratification in Science*. Chicago, University of Chicago Press.

Cole, M. and Scribner, S. (1974). *Culture and Thought: A Psychological Introduction*. New York, Wiley.

Collins, Harry (1985). *Changing Order: Replication and Induction in Scientific Practice*. London and Los Angeles, Sage.

Conklin, Harold (1980). *Ethnographic Atlas of Ifugao: A Study of Environment, Culture and Society in Northern Luzon*. London and New Haven, Yale University Press.

Coutouzis, Mickès (1984). *Sociétés et techniques en voie de déplacement*. Thèse de 3° cycle, Université Paris-Dauphine.

Coutouzis, Mickès and Latour, Bruno (1986). 'Pour une sociologie des techniques: le cas du village solaire de Frango-Castello'. *Année Sociologique*, No. 38, pp.113–167.

Dagognet, Francois (1969). *Tableaux et langages de la chimie*. Paris, Le Seuil.

(1984) *Philosophe de l'image*. Paris, Vrin.

Dubos, René (1951). *Louis Pasteur, Freelance of Science*. London, Golmez.

(1953) and Dubos, J. *The White Plague: Tuberculosis, Man, and Society*, Boston, Little Brown and Co.

Dauben, J. W. (1979). *Georges Cantor: His Mathematics and Philosophy of the Infinite*. Cambridge, Mass., Harvard University Press.

Desmond, Adrian (1975). *The Hot-Blooded Dinosaurs: A Revolution in Paleontology*. London, Blond & Briggs.

Dobbs, Betty, J.T. (1976). *The Foundations of Newton's Alchemy or 'The Hunting of the Greene Lyon'*. Cambridge, Cambridge University Press.

Drake, Stillman (1970). *Galileo Studies: Personality, Tradition and Revolution*. Ann Arbor, University of Michigan Press.

(1978). *Galileo at Work: His Scientific Biography*. Chicago, Chicago University Press.

Duclaux, Emile (1896). *Pasteur: Histoire d'un Esprit*. Sceaux, Charaire.

Easlea, Brian (1980). *Witch-Hunting, Magic and the New Philosophy: An Introduction to the Debates of the Scientific Revolution*. Brighton, Sussex, Harvester Press.

Eisenstein, Elizabeth (1979). *The Printing Press as an Agent of Change*. Cambridge, Cambridge University Press.

Elzen, Boelie (1986). 'The ultracentrifuge: interpretive flexibility and the development of a technological artefact'. *Social studies of science* (forthcoming)

Evans-Pritchard, E.E. (1937/1972). *Witchcraft, Oracles and Magic Among the Azande* (translated from the French). Oxford Clarendon Press.

Fabian, J. (1983). *Time and the Other. How Anthropology Makes its Object*, New York, Columbia University Press.

Farley, J. and J. Geison (1979). 'Science, Politics and Spontaneous generation in 19th century France, the Pasteur–Pouchet Debate', *Bulletin of the History of Medicine*,

Vol. 48, No. 2, pp. 161–198.

Ferguson, Eugene (1977). 'The mind's eye: Nonverbal thought in technology'. *Science*, vol. 197, pp. 827–836.

Foucault, Michel (1975). *Discipline and Punish: The Birth of the Prison* (translated by A. Sheridan). New York, Pantheon.

Fourquet, Francois (1980). *Les Comptes de la puissance*. Paris, Encres.

Freeman, Derek (1983). *Margaret Mead and Samoa: The Making and Unmaking of an Anthropological Myth*. Cambridge, Mass., Harvard University Press.

Garfield, Eugene (1979). *Citation Indexing: Its Theory and Application in Science, Technology and Humanity*. New York, Wiley.

Geison, J. (1974) 'Pasteur' in *Dictionary of Scientific Biography*, 11: 351–415, New York, Scribners & Son.

Gille B. (1978) *Histoire des Techniques*, Paris Gallimard, Bibliothèque de la Pléïade.

Goody, Jack (1977). *The Domestication of the Savage Mind*. Cambridge, Cambridge University Press.

Greimas, A.J. and Courtès, J. (1979/1983). *Semiotic and Language an Analytical Dictionary*. Bloomington, Indiana University Press.

Gusfield, Joseph R. (1981). *The Culture of Public Problems: Drinking-driving and the Symbolic Order*. Chicago. University of Chicago Press.

Hindess, B. (1986). *"Interests' in political analysis'* in J. Law (ed.) pp. 112–131.

Hoddeson, Lilian (1981). 'The emergence of basic research in the Bell telephone system, 1875–1915'. *Technology and Culture*, vol. 22, no. 3, pp. 512–45.

Hollis, M. and S. Lukes (eds.) (1982) *Rationality and Relativism*, Oxford, Blackwell.

Horton, R. (1967). 'African traditional thought and Western science' (complete version). *Africa*, vol. 38 no. 1 pp. 50–71 and no. 2, pp. 155–87.

(1982). 'Tradition and modernity revisited', in M. Hollis and G. Lukes (eds), pp. 201–60.

Hounshell, David A. (1975). 'Elisha Gray and the telephone or the disadvantage of being an expert'. *Technology and Culture*, vol. 6, pp. 133–161.

Hughes, T.P. (1971) *Elmer Sperry: Inventor and Engineer*. Baltimore, Johns Hopkins University Press.

(1979). 'The electrification of America: The system builders'. *Technology and Culture*, vol. 20, no. 1, pp. 124–62.

(1983). *Networks of Power: Electric Supply Systems in the US, England and Germany, 1880–1930*. Baltimore, Johns Hopkins University Press.

Hunter, P. (1980). 'The national system of scientific measurement'. *Science*, vol. 210, pp. 869–74.

Hutchins, E. (1980). *Culture and Inference: A Trobriand Case Study*. Cambridge, Mass., Harvard University Press.

Jenkins, R. (1975). 'Technology and the market: Georges Eastman and the origins of mass amateur photography'. *Technology and Culture*, vol. 15 pp. 1–19.

Kevles, Daniel J. (1985). *In the Name of Eugenics: Genetics and the Use of Human Heredity*. New York, Knopf.

Kevles, David J. (1978). *The Physicists: The History of a Scientific Community in Modern America*. New York, Knopf.

Kidder, Tracy (1981). *The Soul of a New Machine*. London, Allen Lane.

Knorr, Karin (1981). *The Manufacture of Knowledge: An Essay on the Constructivist and Contextual Nature of Science*. Oxford, Pergamon Press.

Knorr, Karin, Krohn, Roger and Whitley, Richard (eds) (1981). *The Social Process of Scientific Investigation*. Dordrecht, Reidel.

Knorr, Karin and Mulkay, Michael (eds) (1983). *Science Observed: Perspectives on the Social Study of Science*. London and Los Angeles, Sage.

Koyré, Alexandre (1966/1978). *Galileo Studies* (translated from the French by J. Mepham). Atlantic Highlands, Humanities Press.

Kuhn, Thomas (1962). *The Structure of Scientific Revolutions.* Chicago, University of Chicago Press.

Lapérouse, Jean-Francois (no date). *Voyages autour du monde.* Paris, Michel de l'Ormeraie.

Latour, Bruno and De Noblet, Jocelyn (eds) (1985). *Les Vues de l'esprit: visualisation et connaissance scientifique. Culture Technique,* numéro 14.

Law, John (1986). 'On the methods of long-distance control: vessels, navigation and the Portuguese route to India'. in J. Law (ed) pp. 234–63.

Law, John (ed.) (1986). *Power, Action and Belief: A New Sociology of Knowledge?* Sociological Review Monograph no. 32 (University of Keele). London, Routledge & Kegan Paul.

Leroi-Gourhan, André (1964). *Le Geste et la parole,* vols 1 and 2. Paris, Albin Michel.

Livingston, Eric (1985). *The Ethnomethodological Foundations of Mathematics* (Studies in Ethnomethodology). London, Routledge & Kegan Paul.

Luria, A.R. (texts edited by M. Cole) (1976). *Cognitive Development: Its Cultural and Social Foundations.* Cambridge, Mass., Harvard University Press.

Lynch, Michael (1985). *Art and Artifact in Laboratory Science: A Study of Shop Work and Shop Talk in a Research Laboratory.* London, Routledge & Kegan Paul.

MacKenzie, D.A. (1978). 'Statistical theory and social interests: a case study'. *Social Studies of Science,* vol. 8, pp. 35–83.

(1981). *Statistics in Britain, 1865–1930.* Edinburgh, Edinburgh University Press.

MacKenzie, D.A. and J. Wajcman, (eds) (1985). *The Social Shaping of Technology.* Milton Keynes, Open University Press.

MacRoberts M.H. and MacRoberts B.R. (1986) Quantitative measures of communication in science: a study of the formal level' *Social Studies of Science,* vol. 16, pp. 151–172.

McNeill, William (1982). *The Pursuit of Power Technology: Armed Forces and Society Since A.D. 1000.* Chicago, University of Chicago Press.

Mead, Margaret (1928). *Coming of Age in Samoa: A Psychological Study of Primitive Youth for Western Civilization.* New York, William Morrow.

Mendelsohn, Everett and Elkana, Yehuda (1981). *Sciences and Cultures (Sociology of the Sciences: A Yearbook).* Dordrecht, Reidel.

Merton, R.K. (1973). *The Sociology of Science: Theoretical and Empirical Investigations.* Chicago, University of Chicago Press.

National Science Foundation (various dates). *Science Indicators.* Washington, DC. NSF.

Nye Mary-Jo (1980) 'N-Rays: An Episode in the History and Psychology of Science' *Historical Studies in the Physical Sciences* vo. 11, pp. 125–156.

(1986). (Science in the Province. Scientific Communities, and Provincial Leadership in France.) California University Press, Berkeley.

Organisation for Economic Co-operation and Development (1984). *Indicators of Science and Technology,* Paris, O.E.C.D. Press.

Perelman, C. (1982). *The Realm of Rhetoric* (translated by W. Kluback). Notre Dame, Indiana, University of Notre Dame Press.

Peters, Thomas and Austin, Nancy (1985). *A Passion for Excellence.* New York, Random House.

Pinch, Trevor (1986). *Confronting Nature: The Sociology of Solar Neutrino Detection.* Dordrecht, Reidel.

Polyani, Michael (1974). *Personal Knowledge: Towards a Post-Critical Philosophy,* Chicago, University of Chicago Press.

Porter, Roy (1977). *The Making of Geology: Earth Science in Britain 1660–1815.* Cambridge, Cambridge University Press.

(1982). 'Charles Lyell: The public and private faces of science'. *Janus*, vol. LXIX, pp.29–50.

Price, Derek de Solla (1975). *Science Since Babylon*. New Haven, Conn., Yale University Press.

Pyenson, Lewis (1985). *Cultural Imperialism and Exact Sciences*. New York, Peter Lang.

Rescher, Nicholas (1978). *Scientific Progress: A Philosophical Essay on the Economis of Research in Natural Science*. Oxford, Blackwell.

Rozenkranz, Barbara (1972). *Public Health in the State, Changing Views in Massachusetts, 1862–1936,* Harvard University Press.

Shapin, Steve (1979). 'The politics of observation: cerebral anatomy and social interests in the Edinburgh phrenology disputes'. in R. Wallis (ed). pp. 139–78.

(1982). 'History of science and its sociological reconstruction'. *History of Science*, vol. 20, pp. 157–211.

Stengers, Isabelle (1983). *Etats et Processus*. Thèse de Doctorat, Université Libre de Bruxelles.

Stevens, Peter S. (1978). *Patterns in Nature*. Boston, Little Brown.

Stocking, G.W. (ed.) (1983). *Observers Observed: Essays on Ethnographic Fieldwork*. Madison, University of Wisconsin Press.

Stokes, T.D. (1982). 'The double-helix and the warped zipper: an exemplary tale'. *Social Studies of Science*, vol. 12, no. 3, pp. 207–40.

Szilard, Leo (ed. S. Weart and G. Szilard) (1978). *Leo Szilard: His Verson of the Facts: Selected Recollections and Correspondence*. Cambridge, Mass., MIT Press

Tolstoy, Leo (1869/1983). *War and Peace* (translated from the Russian by R. Edmunds). Harmondsworth, Penguin.

UNESCO (1983). *Statistical Yearbooks*. Paris, UNESCO.

Wade, Nicholas (1981). *The Nobel Duel*. New York, Anchor Press.

Wallis, Roy (1979). *On the Margins of Science: The Social Construction of Rejected Knowledge*. Sociological Review Monograph, no. 27 (University of Keele). London, Routledge & Kegan Paul.

Watkins, D. (1984). *The English Revolution in Social Medicine 1889–1911*, London, PhD Thesis, University of London.

Wilson, B. (ed) (1970) *Rationality,* Oxford, Blackwell.

Watson, James (1968). *The Double Helix*. New York, Mentor Books.

Wolfe, Tom (1979/1983). *The Right Stuff*. New York, Bantam Books.

Woolgar, Steve (1981). 'Interests and explanations in the social study of science'. *Social Studies of Science*, vol. 11, no. 3, pp. 365–97.

Index

271